WALL ELEMENTS
COLORFUL WALL

建筑立面美化语言
色彩墙

王广福 编著

江苏科学技术出版社

图书在版编目（CIP）数据

建筑立面美化语言：色彩墙 / 王广福编著. -- 南京：江苏科学技术出版社，2013.8
ISBN 978-7-5537-1468-4

Ⅰ.①建… Ⅱ.①王… Ⅲ.①装饰墙—室内色彩—立面造型—世界—图集 Ⅳ.①TU227-64

中国版本图书馆CIP数据核字（2013）第142214号

建筑立面美化语言——色彩墙

编　　著	王广福
项目策划	凤凰空间 / 郑亚男
责任编辑	刘屹立
特约编辑	田　静　李　瑶
责任监制	刘　钧

出版发行	凤凰出版传媒股份有限公司
	江苏科学技术出版社
出版社地址	南京市湖南路1号A楼，邮编：210009
出版社网址	http://www.pspress.cn
总　经　销	天津凤凰空间文化传媒有限公司
总经销网址	http://www.ifengspace.cn
经　　销	全国新华书店
印　　刷	深圳当纳利印刷有限公司

开　　本	965 mm × 1270 mm　1 / 16
印　　张	26
字　　数	208 000
版　　次	2013年8月第1版
印　　次	2013年8月第1次印刷

标准书号	ISBN 978-7-5537-1468-4
定　　价	398.00元（USD 73.00）（精）

图书如有印装质量问题，可随时向销售部调换（电话：022-87893668）。

色彩的释放

文／郑亚男　王青云

颜色是直接影响灵魂的有生力量。

——瓦西里·坎丁斯基

生命是张没价值的白纸，自从绿给了我发展，红给了我热情，黄教我以忠义，蓝教我以高洁，粉红赐我以希望，灰白赠我以悲哀；再完成这帧彩图，黑还要加我以死。从此以后，我便溺爱于我的生命，因为我爱他的色彩。

——闻一多《色彩》

西方人热爱色彩，赋予它人的性情，并且将它涂抹于建筑的结构间，挥洒于日常的生活起居中。徜徉在罗马古城庞贝遗址，在斑驳的镶嵌画中，可臆想罗马贵族们对于颜色的迷恋；置身于法国的夏特尔大教堂中，光透过巨大的玫瑰花窗洒向渺小的祈祷者，能瞬间感受到神的爱抚；在西班牙的圣家族大教堂中，色彩斑驳的石片附着在蜿蜒、婆娑的建筑结构上，奇幻、旖旎不可言表。

西方人热情的特质让他们对色彩的运用大胆而浓艳，而在我们这个东方古国，自古用色就是含蓄而婉约的，从古代诗词里描绘色彩的文字，或者仅仅从命名色彩的文字上，便寻得见那份忧郁和回味，比如：月白、竹青、妃红、胭脂、秋香、黛、檀、绾……那些调和出的美丽柔和的色彩符合东方温润君子及芊芊美人的特质，像极了东方的水墨淡彩。而西方绵延下来的正是他们的油画重彩，这也正是东西方文化和审美情趣的差异在色彩中的体现吧。

当然东西方也有大一统的时候，那就是现代主义的"灰色"风暴。秉承"功能决定形式"这一原则，现代主义建筑在严谨理性的德国人的推动下走向了极致。人们常说，现代主义改变了世界大都会三分之二的天际线，有没有意到现代主义同时还改变了全球三分之二城市的色彩？单调的混凝土立面和玻璃幕墙曾经一统世界，缺少色彩和细节的建筑多少让这个世界变得寡味。当功能与效率不再成为空间利用的首要问题后，人们试图通过更为多样的语言改变整齐划一的节奏，于是"后现代"的时代被开启了。

在这个改造"枯燥"的过程中，人们发现一个真谛，改变建筑的色彩是改变建筑环境最廉价、最简单的途径，尤其对设计枯燥或机构气息浓郁的建筑。适当的颜色方法可帮助设计师构思出和谐的、亲切的及生机勃勃的人工环境，可以与室外的树木、水土和天空构成的自然环境相媲美。

王广福，设计师

毕业于中央美术学院。中艺建筑设计研究院平面与品牌设计研究中心主任。中央美术学院教师。中央美术学院继续教育学院平面设计专业出版与读物方向负责人。曾负责政府、央企、外企、画廊及艺术中心、国家及国际级基金及慈善组织的视觉类设计及画册设计工作。作品曾获第十届全国美展铜奖、深圳03设计优秀奖。

Wang Guangfu, Graphic Designer

Graduated from China Central Academy of Fine Arts (CAFA).
As a director at Brand Research Center of CADI.
Teaching in Graphic Design Department of Continuing Education School of CAFA.

PREFACE

《建筑立面美化语言——色彩墙》就是这样一本介绍世界各地区运用色彩美化建筑的精美案例图集。不同材质、面积、性格、感觉的彩色立面和装置被运用于各类建筑中，与其内在结构、功能定位及地区民俗相得益彰。缤纷的色彩非但没有扰乱人们的视听，反而如同一张得体的名片彰显着建筑的文化内涵与市民的审美品位。

继热销图书《像素墙》与《艺术墙》之后，凤凰空间诚邀中央美术学院教师王广福，共同推出了该系列的第三本书《色彩墙》。与前两本书相比，除了具备前两本海量信息的特点，《色彩墙》图片的绚丽特征让此书更像是一部建筑的抒情诗，细致到每页的色彩分析更具专业性。真心祝愿这本《色彩墙》为国内外的建筑业内人士以及建筑、艺术爱好者喜爱，让它不仅成为您工作时的得力助手，还是您休闲时的枕边益友。

BURST OF COLORS

By Nancy Zheng and Qingyun Wang

"Color is a means of exerting direct influence on the soul."
- Wassily Kandinsky

Life is a piece of blank paper before green gives me development, red gives me enthusiasm, blue gives me nobility, yellow gives me loyalty, blue gives me nobility, pink gives me hope, and gray and white give me sadness. When this color painting is finished, black gives me death. Since then, I have been enamored of my life because I fall in love with its colors.
- Poem **Color** by Wen Yiduo

Westerners are fond of colors and endow them with human emotions. They integrate colors with building structures, and use them in daily living space. Strolling in the ruins of ancient Roman city of Pompeii, we can imagine the Roman nobles' infatuation with colors through the mottled mosaics; standing in the Chartres Cathedral in France, where the sun shines onto the prayers through giant rose windows, we can instantly feel the caress of God; in the Sagrada Familia Church in Spain, the serpentine structure is coated with variegated stone chippings, and it appears magical and magnificent beyond all words.

Westerners' enthusiasm temperament determines that they like to use bold and gaudy colors. In contrast, in China, the oriental country with a long history, people have been used to be implicit and euphemistic in choosing colors since ancient times. It is easy to find the dulcet melancholy and aftertaste in the words describing colors in ancient poetry, or simply the names of the colors. These beautiful and soft colors respond to Eastern men's gentleness and women's demureness. They are very much like the ink and wash painting. In the West, however, it is the gaudy colors in oil painting that have been popular for thousands of years. Probably, this is the difference of culture and esthetic sentiment between Eastern and Western showing in colors.

There was a time when the East was consistent with the West, which is called the "grey storm" of modernism. Following the principle of "form follows function", the rigorous Germans pushed the modernism in architecture to an extreme. it's often said that modernism changed two thirds of the skyline of metropolises in the world. Have people realize that it also changed two thirds of the colors of the world? The world was once occupied by monotonous concrete and glass curtain walls, and the buildings lacking colors and details dim the world more or less. When function and efficiency were no longer the primary problem of space use, people began trying to change the uniform rhythm with various architectural languages. Thus, the postmodern era started.

In the process of transforming the "dullness", we can find that the cheapest and easiest way to change the architectural environment is to change the colors of the architecture, especially for the buildings with a monotonous design or strong institution style. Proper color schemes can help designers create harmonious, amiable and vibrant artificial environment, which can compare favorably with the natural environment composed of trees, water, soil and the sky.

Wall Elements: Colorful Wall is a book introducing the excellent architecture projects around the world which are embellished with colors. The colorful façades and installations of different materials, sizes, characteristics and feelings are employed in all kinds of architecture, and they are in accordance with the architecture's inner structures, functions and local customs. The colors won't interfere with people's view. Instead, they are like a decent name card manifesting the cultural connotation of the architecture and the aesthetic taste of local citizens.

Colorful Wall presented by IFENGSPACE together with Wang Guangfu, a teacher from China Central Academy of Fine Arts, is the third book of the "Wall Elements" series after the best sellers **Pixel Wall** and **Art Wall**. Compared with the previous two books, in addition to the continuous feature of massive amounts of information, the gorgeous pictures in **Colorful Wall** make it more like a lyric poem of architecture, and the detail analysis on colors about each project is very professional. We sincerely hope this book can gain the affection of architecture practitioners and arts lovers as a useful tool at work and a work of art.

七彩缤纷 **RAINBOW COLOR** 011		黛之神秘 **BLACK** 207
绿之生机 **GREEN** 099	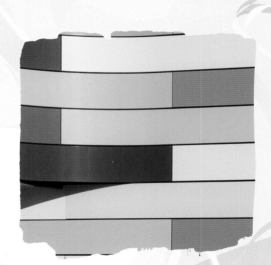	月白之洁 **WHITE** 245
紫之梦幻 **PURPLE** 157		蓝之娴静 **BLUE** 287

赭之怀旧
OCHER
321

赤之热烈
RED
353

WALL ELEMENTS
COLORFUL WALL

七彩缤纷 RAINBOW COLOR

海的幻想 FANTASY OF THE SEA---------12

舞动的森林 THE DANCING FOREST---------16

舞动彩翼的凤凰 PHOENIX'S COLORFUL WINGS---------20

珍宝盒 THE TREASURE BOX---------24

天边的虹彩 THE IRIDESCENCE ON THE HORIZON---------26

巨人的木玩具 GIANTS' WOODEN TOY---------28

七彩翻板游戏 THE GAME OF COLORFUL PLANKS---------34

孩子们的城堡乐园 FUN CASTLE FOR CHILDREN---------38

芬芳的花草地 A FRAGRANT LAND OF GRASS AND FLOWERS---------40

童年的世界 THE WORLD OF CHILDHOOD---------46

风中的芦苇塘 THE REED FIELD IN THE WIND---------50

青蛙王子 FROG KING---------52

彩色条纹的欢快韵律 THE GAILY RHYTHM OF COLORFUL STRIPES---------56

彩虹雨 RAINBOW SHOWER---------58

彩色方块世界 A WORLD OF COLORFUL SQUARES---------62

彩虹千层派 RAINBOW MILLE - FEUILLE---------66

魔幻树叶 MAGIC LEAVES---------70

色彩的王国 A KINGDOM OF COLORS---------74

幻彩殿堂 COLORFUL PALACE---------76

南十字星闪耀 THE SHINNING SOUTHERN CROSS---------78

自然之园 THE KINDERGARTEN OF NATURE---------80

彩虹屋 RAINBOW HOUSE---------84

流动的幻彩韵律 THE FLOWING RHYTHMS---------88

色彩斑斓的住宅 COLORFUL HOUSING---------92

虚幻的蛊惑 PSYCHEDELIC DELUSION---------94

绿之生机 GREEN

生机盎然的森林 VITALITY OF FOREST---------100

双色太阳镜 DUAL COLOR SUNGLASSES---------102

动物的天堂 ANIMALS' HEAVEN---------104

神奇小人国 THE MYSTERIOUS LILLIPUTIAN---------106

绿意盎然 FULL OF GREEN---------108

律动的生机 DYNAMIC VITALITY---------110

七彩童年 COLORFUL CHILDHOOD---------112

悄然绽放的花朵 A SILENTLY BLOOMING FLOWER---------116

杨树林的守卫 POPLAR GROVE'S ESCORT---------120

绿铅笔 THE GREEN PENCILS---------122

绿光森林 GREEN FOREST---------126

目录 CONTENTS

水的世界 THE WORLD OF WATER --------130

绿草艾艾 GREEN GRASS --------134

给予绿色的关怀 GREEN CARE --------136

未成熟的青涩 IMMATURE YOUTH --------140

一道绿色风景线 A GREEN SCENERY --------144

一方洁净的空间 A PIECE OF CLEAN SPACE --------146

雨露滋润、生机如笋
LIKE VIVIFYING BAMBOO SHOOTS BATHED IN RAIN AND DEW --------150

梦想的舞台 STAGE FOR DREAMS --------152

浴火卫士 FIRE GUARD --------154

紫之梦幻 PURPLE

穿越时空的水晶球 A CRYSTAL BALL ACROSS TIME AND SPACE --------158

宝石与梦境的圆舞曲 THE WALTZ OF GEMSTONES AND DREAMS --------160

炫彩万花筒般的奇幻世界 A KALEIDOSCOPIC WORLD --------164

山脚下的梦想城堡
THE DREAM CASTLE AT THE FOOT OF THE MOUNTAIN --------168

历史的展现、唯美的重生
REAPPEARANCE OF HISTORY, REBIRTH OF AESTHETICISM --------172

水晶花瓣 CRYSTAL PETALS --------174

通往星空的时光隧道 THE TIME TUNNEL TO THE STARRY SKY --------176

城海中的小憩与欢愉之岛 A PLEASURE ISLAND IN THE CITY --------180

霓虹光晕 The Halo of Neon Lights --------182

柔光萦绕、花瓣纷飞 SOFT LIGHT HOVERING FLYING PETALS --------186

孤静相随、归之净土 THE SOLITARY LAND TO RETURN TO --------188

功能与色彩的艺术之舞
THE ARTISTIC DANCE OF FUNCTION AND COLOR --------190

舞动色彩、赋予意义 DANCING COLORS, GIVE MEANING --------192

行云流水 SMOOTH LIKE FLOATING CLOUDS AND FLOWING WATER --------194

高雅购物天堂 ELEGANT SHOPPING PARADISE --------198

彩翼 THE COLORFUL WINGS --------200

弦外之音 OVERSTONE --------204

黛之神秘 BLACK

黑白时尚 BLACK AND WHITE FASHION --------208

外表朴实、内有乾坤
A SLIENT BOX WITH WONDERFUL CONTENTS HIDDEN INSIDE --------210

神秘波浪 MYSTERIOUS WAVES --------212

神秘黑色住宅 HOUSING IN MYSTERIOUS BLACK --------214

行走的雕塑 MOVING SCULPTURE --------216

扭曲的黑盒子 TWISTED BLACK BOX --------218

折叠的壳 FOLDED SHELL --------222

迷幻银盒 MAGIC SILVER BOX-----226

为了忘却和理解的纪念
A COMMEMORATION FOR FORGET AND UNDERSTANDING-----228

坚实的守护 STAUNCH PROTECTION-----232

黑暗中的水晶城堡 CRYSTAL CASTLE IN DARKNESS-----236

意象与现实 IMAGE REALITY-----240

月白之洁 WHITE

花开物语 FLOWER IMPLICATION-----246

朦胧之美 HAZY BEAUTY-----248

与风共舞 MOVE WITH THE WIND-----252

银妆素裹 A WHITE WORLD IN SNOW-----254

白色时光 WHITE TIME-----256

纯净的世界 THE WHITE AND PURE WORLD-----260

涟漪轻漾 GENTLE RIPPLES ON THE SURFACE-----264

优雅·通透·轻盈 GRACEFUL·CRYSTAL·ETHEREAL-----268

在云端 IN THE CLOUDS-----270

轻盈之美 THE BEAUTY OF LIGHTNESS-----274

光的游乐场 HOUSE OF LIGHT-----276

流动的音符 FLOWING MUSIC-----280

亲近自然 CLOSE TO NATURE-----282

蓝之娴静 BLUE

精灵之家 ELF HOME-----288

悦动流转的五线谱 FLOWING STAVE-----292

世外桃源 A FICTITIOUS LAND OF PEACE-----296

温馨的回忆 SWEET MEMORIES-----300

流动变幻的光芒之旅 THE JOURNEY OF FLOWING LIGHT-----302

天边的尖角城堡 REMOTE CASTLE WITH SHARP ANGLES-----304

发光的树 TREES OF LIGHT-----306

希望之光 LIGHT OF HOPE-----310

诡秘几何 MYSTERIOUS GEOMETRY-----312

蓝色向往 THE BLUE DREAM-----314

太阳能生态住宅 SOLAR RESIDENCE-----316

奢华盛宴 LUXURY FEAST-----318

赭之怀旧 OCHER

花园中轻摇的风铃
AEOLIAN BELLS GENTLY SWINGING IN THE GARDEN-----322

锈蚀的回忆 RUSTY MEMORIES-----326

缤纷春季乐园 COLORFUL SPRING FAIRYLAND-----330

目录 CONTENTS

岁月的沉淀 SEDIMENT OF TIME----------334

穿越古老的记忆 TRARELING THROUGH THE OLD MEMORIES------336

蒲公英的世界 A WORLD OF DANDELION----------340

返璞归真 RETURNING TO THE ORIGINAL NATURE----------344

成熟的果实 RIPE FRUIT----------346

沁人心脾的温暖 THE SUNSHINE WARMTH BRINGS JOY TO THE SOUL--350

赤之热烈 RED

逐光追影 RED COLOR DANCE WITH LIGHT AND SHADOW--------354

热烈的韵律 ARDENT RHYTHMS----------358

诗意的栖居 POETIC DWELLING----------360

美妙童梦 A MAGICAL DREAM FOR CHILDREN----------362

活力橙色空间 DYNAMIC ORANGE SPACE----------366

一米阳光 A RAY OF SUNSHINE----------368

飘香热咖啡 THE WAVES OF HOT COFFEE----------370

万花筒般的世界 A KALEIDOSCOPIC WORLD----------372

青春飞扬活力四射 VIGOR OF YOUTH----------378

为生活空间增添活力
IMPROVING THE LIVING SPACE WITH VITALITY----------380

魅惑夜色 GLAMOROUS NIGHT----------382

迷幻之夜 PSYCHEDELIC NIGHT----------386

吉卜赛韵味 GYPSY FLAVOR----------390

热辣魅惑的空间 A HOT CHARM OF SPACE----------392

成长的力量 THE POWER OF GROWTH----------394

奇妙之地 A PLACE OF WONDERS----------396

红与白融合激情与浪漫
RED AND WHITE BLEND PASSION AND ROMANCE----------398

秋天的童话 AUTUMN FAIRYTALE----------400

活力无限 ENDLESS FLOW OF ENERGY----------404

发现之旅 DISCOVERY TRAVEL----------406

架起沟通的桥梁
BRIDGING THE COMMUNICATION GAP BETWEEN PEOPLE----------408

木条搭建的奇幻世界
THE WOODEN STRUCTURES FANTASY WORLD----------410

索引 INDEX----------412

■七彩,赤、橙、黄、绿、青、蓝、紫,古人以这突出的七种颜色,来指代绚丽多彩的图像。■全色系色彩灵活运用,构成了缤纷的建筑立面,轻松地让建筑物脱颖而出。■全色系的色彩运用是美丽的、高调的,多用于公共空间或幼儿园、学校等建筑。在现代建筑中运用的尤为精彩。■全色系的运用是最需慎重的,因为环境中太多明亮的色彩会产生负面的效果——使人过度兴奋、躁动、烦躁和紧张。降低全色系色彩的明度和纯度,减少全色系颜色的面积,以及加入大面积的中间色调和都不失为一种好的解决办法。

■ Rainbow color, including red, orange, yellow, green, blue, cyan and purple, is used to describe the colorful picture since ancient time. ■ Flexible use the rainbow color will create colorful façades and help architecture stand out. ■ The rainbow color is high-profile, and mostly used for public spaces, kindergartens, schools, etc. And its application is especially splendid in modern architecture. ■ Rainbow color should be used with caution because too many bright colors in the environment will bring negative effects, such as making people overexcited, restless, agitated and nervous. A good solution is to reduce the brightness, purity, proportion of these colors and add large areas of intermediate colors.

七彩缤纷
RAINBOW COLOR
011

海的幻想
FANTASY OF THE SEA

纯黄 Yellow

橙色 Orange

深橙色 Dark Orange

月白 Moon White

亮天蓝色 Light Sky Blue

EL "B" 礼堂，Selgascano，西班牙，卡塔赫纳，2011年

EL "B" 礼堂的造型强调体量感，用不同的色彩营造简洁的体块，表面平整连续，切割堆叠变化有序。以条状半透明的塑料（具有防紫外线功能的烯酸甲酯）作为建筑装饰的主材，拼装形成许多大的色块。但具体到每个单独的色块，色条颜色丰富，柔和又有微差，远远望去，如梦如幻。

EL "B" Auditorium, Selgascano, Cartagena, Spain, 2011

EL "B" Auditorium emphasizes a sense of volume. It has a simple shape and a smooth, continuous surface, and it is orderly-cut and arranged. The main material of the façade is translucent plastic, which are mostly in strips. They are perfectly organized to create numerous lines, and the colors are all smooth and soft, but slightly different. From afar, the building appears a little unrealistic because of the optical illusions.

材料：铝、塑料（具有防紫外线功能的烯酸甲酯）

Materials: aluminum, plastic (ultraviolet-proof polymethyl methacrylate)

东立面图 EAST ELEVATION

西立面图 WEST ELEVATION

北立面图 NORTH ELEVATION

南立面图 SOUTH ELEVATION

© Iwan Baan

© Iwan Baan

© Iwan Baan

爱丽丝蓝 Alice Blue

 茶色 Tan

 橙色 Orange

 深橙色 Dark Orange

 橙红色 Orange Red

 黄绿色 YellowGreen

 中兰花紫 Medium Orchid

 暗灰蓝色 Dark Slate Blue

© Iwan Baan

© Iwan Baan

EL "B" 礼堂，SelgasCano，西班牙，卡塔赫纳，2011年

　　EL "B" 礼堂的室内更是使用了一组海洋气息浓郁的色彩，立面的蓝色来自海洋，而塑料条（具有防火功能的透明和半透明的聚碳酸酯）带来海洋面的波光粼粼之感。顶面的橘红来自海上的夕阳，地板色浅，恍若白色沙滩，七彩的座椅点缀其间，一组温暖、干净、积极向上的色彩，完美了诠释了对"海的幻想"。

EL "B" Auditorium, SelgasCano, Cartagena, Spain, 2011

The interior color scheme is closely related to the ocean. The color of the sea was used on the façade and the plastic stripes were originated from the glistening light of waves; the orange ceiling has the color of the setting sun over the sea; the light-colored floor is just like the white beach. The colorful chairs are interspersed among. The "Fantasy of the Sea" was perfectly interpreted through a group of warm, clean and lively colors.

材料：铝、塑料（具有防紫外线功能的烯酸甲酯）

Materials: aluminum, plastic (ultraviolet-proof polymethyl methacrylate)

© Iwan Baan

 暗绿宝石 Dark Turquoise

 深天蓝 Deep Sky Blue

 橙色 Orange

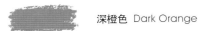 深橙色 Dark Orange

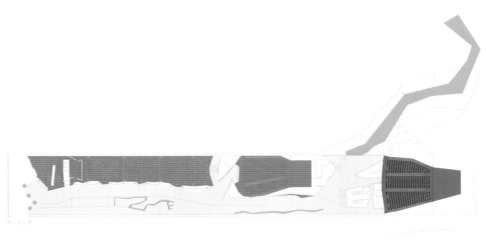

下层平面图 LOWER FLOOR PLAN

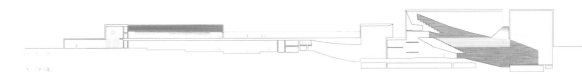

纵向剖面图 LONGITUDINAL SECTION

© Iwan Baan

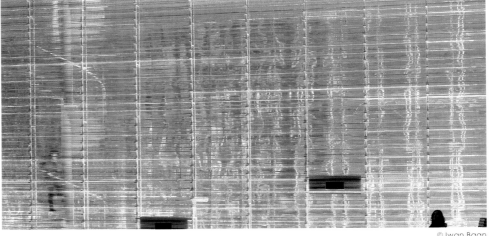

© Iwan Baan

七彩缤纷 | RAINBOW COLOR

舞动的森林
THE DANCING FOREST

 栗色 Maroon

 橙红色 Orange Red

 纯黄 Yellow

 黄绿色 Yellow Green

 矢车菊蓝 Cornflower Blue

 钢蓝 Steel Blue

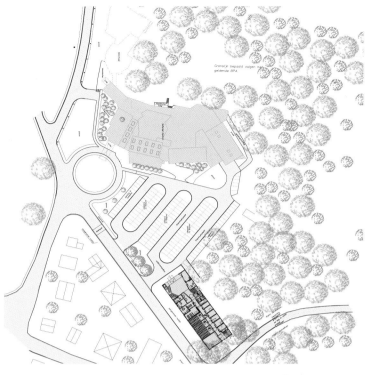

总平面图 SITE PLAN

立面图 ELEVATION

© Miguel de Guzmán

MWD学院，Carlos Arroyo Arquitectos，比利时，迪尔贝克，2012年

建筑的外表皮十分有趣，当你向林区方向走过去，条形表皮映射出树林的形象，当你向相反方向走去，看到是多彩的条形立面，步移景异，变幻无穷。建筑外表皮是身兼建筑师和色彩专家的Alfons Hoppenbrouwers的研究成果，色彩经过他的数字化的组合，比例，节奏，长度，形式，纹理，颜色各方面具有奇妙的节奏感和韵律感，就像音乐般和谐动人、变幻莫测，展现出建筑、天空、森林多涵义的美。

Academie MWD, Carlos Arroyo Arquitectos, Dilbeek, Belgium, 2012

The skin of the building is pretty interesting. If you walk towards the reserve forest, the skin reflects an image of a forest. If you walk in the opposite direction, it is a façade composed of colorful stripes. The scenery changes with your movement. The skin is the research achievement of Alfons Hoppenbrouwers, an architect and color expert. The façade is a combination of mathematics and color. A rhythm is reveled through proportion, pattern, length, form, texture and color. It is as harmonious, pleasant and varying as music, interpreting the beauty of the building, the sky and the forest.

材料：铝、塑料（具有防火功能的透明和半透明的聚碳酸酯）、涂料、木材

Materials: aluminum, plastic (transparent and translucent fire-proof polycarbonate), coating, wood

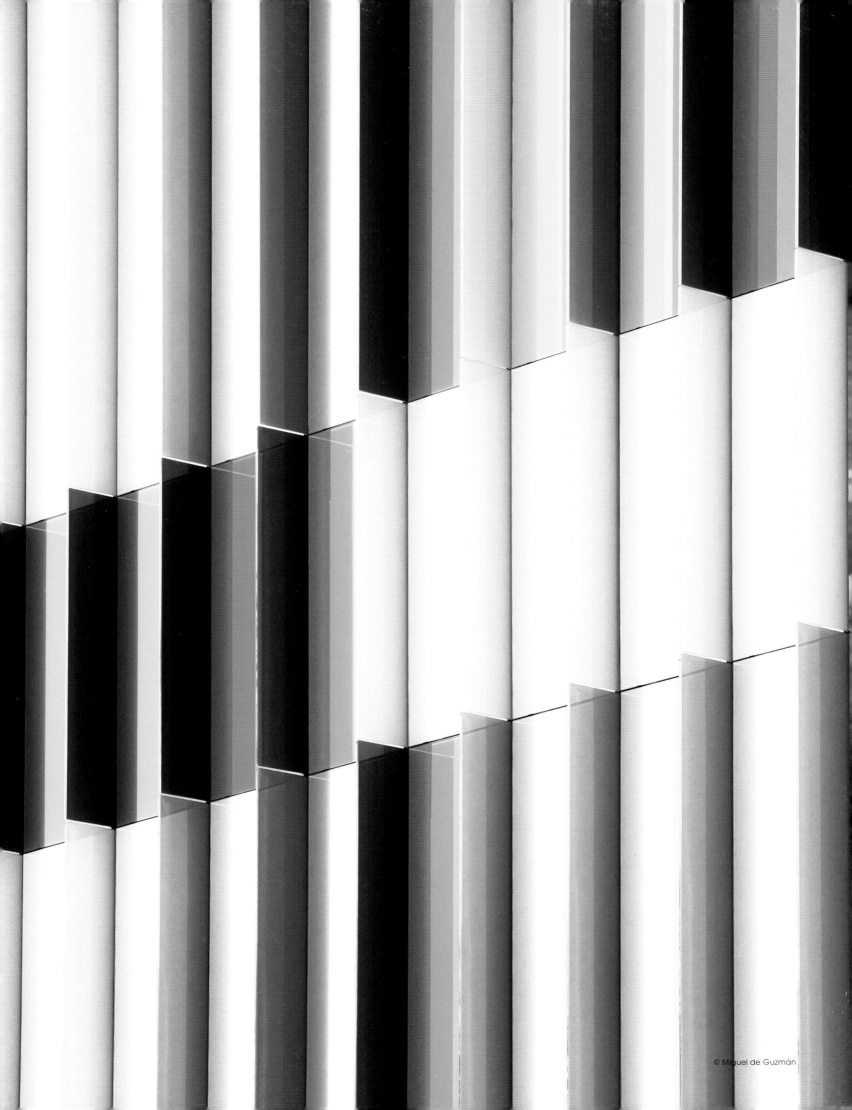

© Miguel de Guzmán

栗色 Maroon

 橙红色 Orange Red

 纯黄 Yellow

 黄绿色 Yellow Green

 矢车菊蓝 Cornflower Blue

 钢蓝 Steel Blue

© Miguel de Guzmán

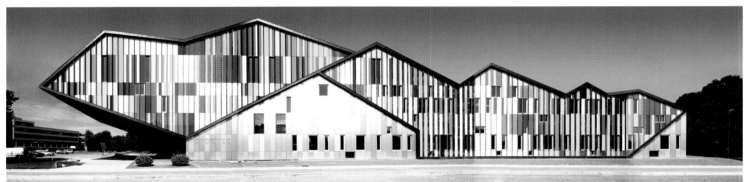
© Miguel de Guzmán

© Miguel de Guzmán

月白 Moon White

 淡灰色 Light Gray

亮蓝灰 Light Slate Gray

MWD学院，Carlos Arroyo Arquitectos，比利时，迪尔贝克，2012年
　　建筑的背面虽然也是色块的拼接，但相对正面的繁茂的色彩和变换无穷的效果，背面只采用素净的灰白，窗口隐匿其间，旨在让适量的间接阳光穿过。建筑场地限制诸多，南靠市政广场和餐厅，西边是Westrand文化中心，北侧是天然林保护区，东边则是尖屋顶的密集郊区别墅区。如何协调场地环境差异性并突显自己的特色？不同立面色彩、纹理、材质的使用起到了关键的作用。

Academie MWD, Carlos Arroyo Arquitectos, Dilbeek, Belgium, 2012
Although the back of the building is also a combination of color blocks, compared with the rich colors and varying effects of the façade, only white and gray are used. Windows are hidden inside to introduce a proper amount of indirect light. The building is in a difficult context with a variety of contrasting situations: south, the main square (Gemeenteplein) with the City Hall and local restaurants; west, CC Westrand; north, a protected area of natural forest; and east, a compact group of suburban villas with pitched roofs. How to coordinate the environment variety and emphasize its own style? Different colors, textures and materials play a prominent role.

材料：金属板、涂料、玻璃

Materials: metal plate, coating, glass

一层平面图 GROUND FLOOR PLAN

剖面图 SECTION

© Miguel de Guzmán

舞动彩翼的凤凰
PHOENIX'S COLORFUL WINGS

 桃肉色 Peach Puff

 中海洋绿 Medium Sea Green

 深洋红 Dark Magenta

 黛青色 Dark Cyan

紫兰色 Indigo

雅乐轩伦敦埃克塞尔酒店，Jestico + Whiles，英国，伦敦，2011年

　　受Bridget Riley 某些绘画作品中立体效果的启发，建筑师们为雅乐轩设计了熔结在玻璃中的抽象横纹和斑斓的色彩，着重强调了建筑物的流动性。酒店呈"蛇"形布局，凹凸起伏，呈现出流动性和立体感。另外，立面颜色随着一天内光照的变化，呈现自然的色彩渐变。远远望去，一片姹紫嫣红，满眼色彩斑斓，令人赏心悦目。

Aloft London Excel, Jestico + Whiles, London, UK, 2011

Inspired by the illusion of depth seen in some of Bridget Riley's paintings, abstract horizontal stripes fritted onto the glass have been designed to emphasize the flow of the building. This creates an unusual interplay of movement and depth when combined with the undulating colors of the crescent wings. Aloft London Excel is in a serpentine shape with undulating convex concave rhythm, creating a sense of fluidity and stereo perception. Besides, the color of the façade changes gradually due to the daylight. Seen from afar, the building is like a blaze of colors, a feast for the eyes.

材料：反光金属板、陶板材、玻璃

Materials: reflective metal plates, ceramic tiles, glass

© Tim Crocker

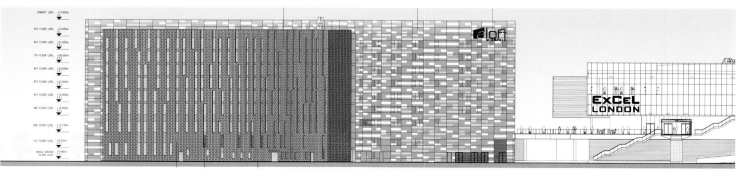
北立面图 NORTH ELEVATION

© Tim Crocker

淡灰色 Gainsboro

深灰色 Dark Gray

纯红 Red

黛青色 Dark Cyan

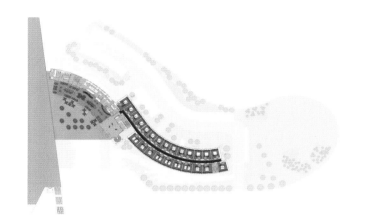

平面图 PLANS

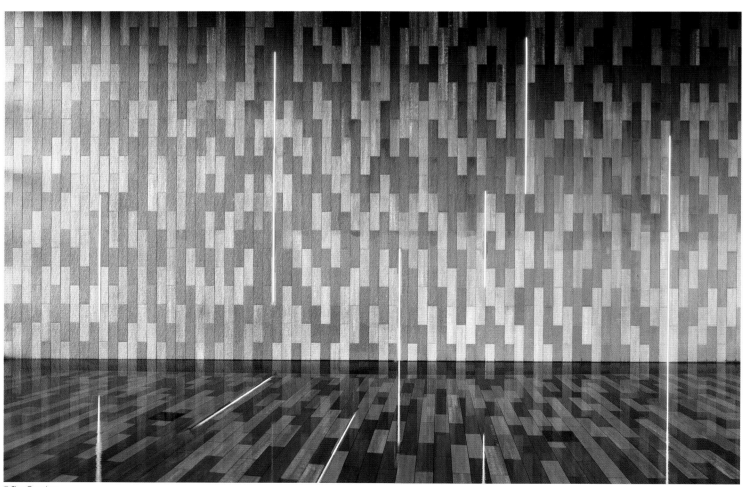

© Tim Crocker

© Tim Crocker

剖面图 SECTION

雅乐轩伦敦埃克塞尔酒店,Jestico + Whiles,英国,伦敦,2011年

在内部空间的处理上,设计师延续了外部的多色线形设计。创新型的材料和定制的彩色透光板装饰着客房走廊,使人不禁联想起立面的设计,从而增强了内外空间的衔接性。那些漂亮的木质材料、人造大理石、玻璃制品、瓷器、有织纹的墙纸、雕塑和绘画作品的交相辉映,相互协调,好似正在演奏着巴洛克式的华美交响乐。

Aloft London Excel, Jestico + Whiles, London, UK, 2011

The inner space continues the linear, multi-colored motif of the external design. Innovative finishes and bespoke colored light panels, reminiscent of the façade treatment, decorate the guestroom corridors, creating cohesion between the internal and external spaces. The beautiful wood, artificial marble, glass products, ceramics, wall papers with woven stripes, sculptors and paintings enhance each other's beauty and cooperate in harmony, as if a magnificent Baroque movement is being played.

材料:木质材料、人造大理石、玻璃制品、瓷器、有织纹的墙纸、雕塑、绘画作品

Materials: wood, artificial marble, glass products, ceramics, wall papers with woven stripes, sculptors, paintings

珍宝盒
THE TREASURE BOX

 亮天蓝色 Light Sky Blue

 金色 Gold

 橙红色 Orange Red

 硬木色 Burly Wood

布兰德霍斯特博物馆，Sauerbruch Hutton，德国，慕尼黑，2009年

从深紫到淡黄的23种颜色的36 000根陶棒垂直排列于水平方向的金属板上，用螺栓固定于板面上。每根彩色陶棒的尺寸为4 cm×4 cm×110 cm，通过不同的排列方式以及渐变色调把整个表皮分成三大色块，如同用一块经过精心裁减并拼贴而成的花布，包裹住在这个博物馆建筑外面，召唤着人们对里面收藏的宝贝展开一次奇妙的艺术之旅。博物馆的外表皮分为内外两层，里层是直接固定于结构面上的水平向双色穿孔金属板，设计师显然是希望通过这些密布于板上的小孔来吸收噪声，减少紧邻主干道的场地环境对博物馆内安静氛围的干扰。

Brandhorst Museum, Sauerbruch Hutton, Munich, Germany, 2009

The external skin of the building is composed of 36,000 ceramic rods in an assortment of 23 custom colors from dark purple to light yellow. The rods are vertically arranged on a horizontal metal plates and fixed with bolts. All the rods are 4×4×110 cm and are arranged in different ways. The surface is divided into three color blocks by the different arrangements of the rods and a gradual color. The surface wraps the museum like a piece of exquisitely cut and spliced cloth. It attracts people to start a marvelous art trip for the collections inside. The museum has a double layer. The inner layer consists of a horizontally folded metal skin, coated in two colors. The architects intend to absorb noises through the densely arranged holes so as to reduce the disturbance to the tranquil ambience of the museum caused the by the environment which is close to the main road.

材料：4 cm×4 cm×110 cm 陶瓷棒、金属板

Materials: 4 cm×4 cm×110 cm ceramic rods, metal plates

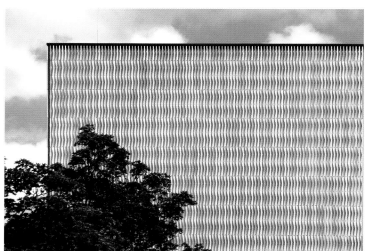

© Courtesy of Sauerbruch Hutton

© Courtesy of Sauerbruch Hutton

© Courtesy of Sauerbruch Hutton

天边的虹彩
THE IRIDESCENCE ON THE HORIZON

粉红 Pink

亮粉红 Light Pink

闪兰色 Dodger Blue

纯蓝 Blue

橙红色 Orange Red

金色 Gold

宝石碧绿 Aquamarine

总平面图 SITE PLAN

© Andrei Margulescu

© Andrei Margulescu

光谱住宅，Re-Act Now，罗马尼亚，康斯坦察，2009年

赤、橙、黄、绿、青、蓝、紫，全色系都用在建筑上又如何？Re-Act Now即在建筑中用了所有的颜色，让整片大楼成为天边的一抹虹彩。远观、近看、身在其中都有一种缤纷的心情，缤纷的世界让一切显得虚幻，产生了幻觉。它是一种精神的视差效果，但是仍然在你的心中，在你的感官、意识和潜意识里。

Spectrum Residential Ensemble, Re-Act Now, Constanta, Romania, 2009

What if red, orange, yellow, green, cyan, blue and purple are all used on a building? Re-Act Now did so and made the building the Iridescence on the horizon. A colorful mood will rise in you whether you see the building closely, from far away or in it. The colorful world makes everything unrealize and produces illusion, cloud and mist, fantasy... A spiritual parallax though it is, it is still in your heart, sense organs, consciousness, and sub-consciousness.

材料：金属面板

Material: metal panels

© Andrei Margulescu

© Andrei Margulescu

巨人的木玩具
GIANTS' WOODEN TOY

淡青色 Light Cyan

纯黄 Yellow

亮天蓝色 Light Sky Blue

纯绿 Green

纯蓝 Blue

猩红 Crimson

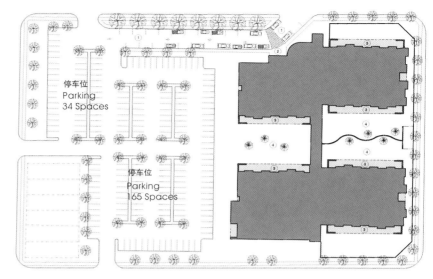

1 落客车道　1 Drop-off Lane
2 入口　　　2 Entry
3 走廊　　　3 Porch
4 游乐区　　4 Play Area

总平面图 SITE PLAN

© Scott McDonald

切萨皮克儿童发展中心，Elliott + Associates Architects，美国，俄克拉何马州，俄克拉何马市，2011年

想象这样一个地方，这里是孩子的专属之地，而成人仅仅来此参观；

想象这样一个地方，这里阳光明媚，蝶鸢摇曳；

想象着彩虹的色彩遍染大地，三轮车道和一方碧水让你穿行而过。

Chesapeake Child Development Center, Elliott + Associates Architects, Oklahoma City, Oklahoma, USA, 2011

Imagine a place made for little people where adults just visit...
Think about a place filled with sunlight, butterflies and kites...
Imagine the colors of the rainbow, a trike track and water you can run through...

材料：砖、金属、玻璃、混凝土

Materials: bricks, metal, glass, concrete

© Scott McDonald

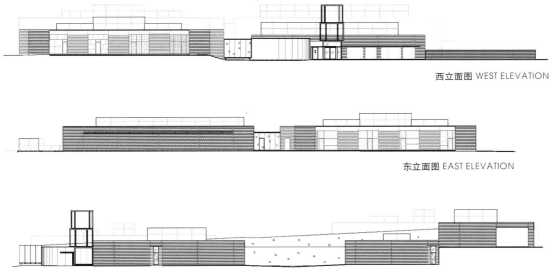

西立面图 WEST ELEVATION

东立面图 EAST ELEVATION

南立面图 SOUTH ELEVATION

北立面图 NORTH ELEVATION

© Scott McDonald

© Scott McDonald

© Scott McDonald

纯黄 Yellow

 沙棕色 Sandy Brown

 深橙色 Dark Orange

 猩红 Crimson

 淡绿色 Light Green

 亮蓝 Light Blue

© Scott McDonald

切萨皮克儿童发展中心,Elliott + Associates Architects,美国,俄克拉何马州,俄克拉何马市,2011年

将有会这样一个地方,让你找寻气球狗和风中旋转的风车;

想象着积木高高堆叠,你可以从字母下穿过,还有植物搭起的大象陪伴着;

这里会有纳凉的大树,有趣的故事和愉快的野炊;

想象一幅幅冠蓝鸦、花朵、南瓜的图画,还有苹果让你抚摸,巨大的文字每天将你迎接。

Chesapeake Child Development Center, Elliott + Associates Architects, Oklahoma City, Oklahoma, USA, 2011

There will be a place to discover a balloon dog and a pinwheel spinning in the wind…

Think about blocks piled high, alphabet letters you can walk under and an elephant made of plants…

There will be a big tree for shade and storytelling and picnics…

Imagine blue jays and flowers and pumpkin pictures and apples you can touch and big words to greet you each day…

材料:吸声天棚、木材、涂料、塑料

Materials: acoustical ceiling, wood, paint, plastic

© Scott McDonald

闪兰色 Dodger Blue

 暗蓝色 Dark Blue

 深橙色 Dark Orange

切萨皮克儿童发展中心，Elliott + Associates Architects，美国，俄克拉何马州，俄克拉何马市，2011年

这是一座教授知识的教学楼，

一块充满了幻想与奇迹的巨大书写板，

一个让想象力翱翔的地方；

请将它想象成彩色图画书的张页，

或是花园中的一件玩具。

Chesapeake Child Development Center, Elliott + Associates Architects, Oklahoma City, Oklahoma, USA, 2011

Our place will be a building that teaches;

a giant storyboard filled with visual fantasy and wonder;

a place to make your imagination soar…

Think of the building as pages of a coloring book…

Think of this place as a toy in the garden…

材料：吸声天棚、木材、涂料、塑料

Materials: acoustical ceiling, wood, paint, plastic

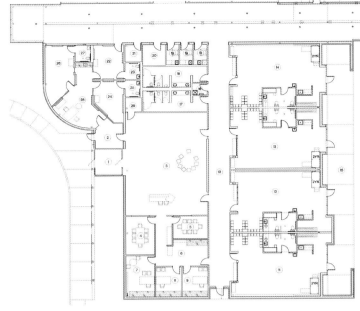

1 前厅	15 走廊	1 Vestibule	15 Porch
2 前厅	16 管理员办公室	2 Vestibule	16 Janitor
3 大堂	17 女卫生间	3 Lobby	17 Women
4 会议室	18 男卫生间	4 Conference	18 Men
5 会议室	19 哺乳室	5 Conference	19 Lactation Room
6 复印室	20 储藏室	6 Copy	20 Storage
7 主管办公室	21 资料室	7 Director	21 Data
8 办公室	22 病房	8 Office	22 Sick Room
9 办公室	23 卫生间	9 Office	23 Toilet
10 走廊	24 病房	10 Corridor	24 Sick Room
11 教室	25 卫生间	11 Classroom	25 Toilet
12 教室	26 病房	12 Classroom	26 Sick Room
13 教室	27 卫生间	13 Classroom	27 Toilet
14 教室	28 看护室	14 Classroom	28 Nurse
	29 咖啡厅		29 Coffee

平面图 PLAN

© Scott McDonald

© Scott McDonald

金色 Gold

 纯绿 Green

 猩红 Crimson

© Scott McDonald

© Scott McDonald

七彩缤纷 | RAINBOW COLOR | 33

七彩翻板游戏
THE GAME OF COLORFUL PLANKS

纯黄 Yellow

橙红色 Orange Red

紫色 Purple

深灰色 Dark Gray

猩红 Crimson

闪兰色 Dodger Blue

闪光深绿 Lime Green

粉红 Pink

亮天蓝色 Light Sky Blue

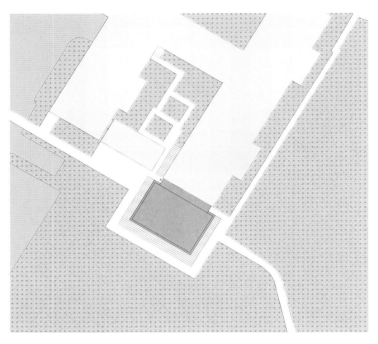

总平面图 SITE PLAN

© Miran Kambič

© Miran Kambič

可可幼儿园，Arhitektura Jure Kotnik，斯洛文尼亚，卢布尔雅那，2010年

建筑成功不仅在于木条上的9种颜色选得好，更在于把建筑做成了孩子们的一个大玩具，与孩子们直接的产生互动。这些木板就如同百叶窗的叶片，可以绕着中央的金属轴转动。木板的一面保留木材的原色，另外一面刷上了9种鲜艳的颜色，拼在一起就像是彩色的琴键。

Kindergarten Kekec, Arhitektura Jure Kotnik, Ljubljana, Slovenia, 2010

The building succeeded not only in selecting the nine fine colors for the planks, but also in being treated as a big toy, which can interact directly with the children. These louver-like planks can revolve around their vertical metal axis. They keep the color of natural wood on one side but painted into nine different bright colors on the other side.

材料：木板、涂料

Materials: planks, paint

© Miran Kambič

© Miran Kambič

© Miran Kambič

翻版拼图板 LAMELLAS PALETTE

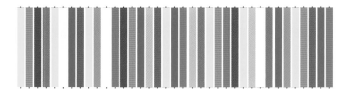
彩虹立面 FAÇADE RAINBOW

原木色和彩色拼法 WOOD VS. PANTION 674 C

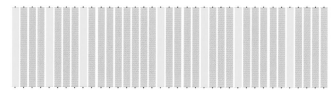
原木色和彩色拼法 WOOD VS. PANTION 394 C

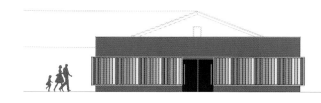

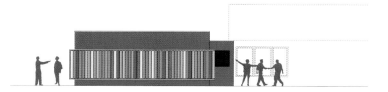

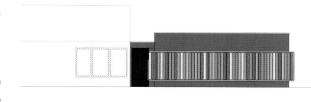

立面图 FAÇADES

可可幼儿园，Arhitektura Jure Kotnik，斯洛文尼亚，卢布尔雅那，2010年

彩色木板墙建成后，首先成为了教学工具，孩子们可以通过它来辨识颜色；其次，它也是孩子们最爱的玩具。孩子们每天都可以通过转动木板给小木屋一个全新面貌。

Kindergarten Kekec, Arhitektura Jure Kotnik, Ljubljana, Slovenia, 2010
The colorful planks firstly became a learning tool for children distinguish different colors; secondly, they are the children's favorite toy. The children give wooden house a new appearance everyday by manipulating colorful wooden planks.

材料：木板、涂料

Materials: planks, paint

秘鲁色 Peru

 弱紫罗兰红 Pale Violet Red

 古董白 Antique White

可可幼儿园，Arhitektura Jure Kotnik，斯洛文尼亚，卢布尔雅那，2010年

室内的颜色相对的温和，狭长的更衣室内里面有一排衣橱，由原木制成，衣柜下的原木鞋柜刷成五颜六色，以便与外墙面形成呼应。这些鞋柜还身兼两职，收进去是鞋柜，抽出来就能当凳子坐。

Kindergarten Kekec, Arhitektura Jure Kotnik, Ljubljana, Slovenia, 2010

The interior color scheme is relatively mild. Wardrobes in the narrow changing room are made from pure natural wood and have pull-out boxes for shoes in all the colors of the façade, which function as a space saver, since they also serve as a bench.

材料：木板、涂料

Materials: planks, paint

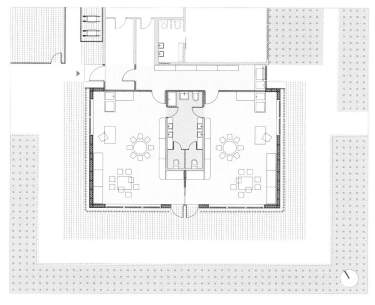

平面图 PLAN

© Miran Kambič

© Miran Kambič

孩子们的城堡乐园
FUN CASTLE FOR CHILDREN

	柠檬绸 Lemon Chiffon
	橙色 Orange
	闪光深绿 Lime Green
	暗绿宝石 Dark Turquoise
	矢车菊蓝 Cornflower Blue
	中蓝色 Medium Blue
	深粉红 Deep Pink

总平面图 SITE PLAN

© Fernando Alda

© Fernando Alda

Los Mondragones日托中心和市政餐厅，Elisa Valero Ramos，西班牙，格拉纳达，2006年

该项目的主旨是将建筑融于既有环境。建筑缩短了外立面，为铺装各种色彩的玻璃纤维增强水泥喷漆板腾出了空间。色彩的使用为原建筑增添了一抹亮彩。高高的窗户朝向河畔绿树隔滤的夏日晨光，西侧混凝土框架的玻璃墙朝向白色的花园。

Daycare Center and Municipal Dining Hall at Los Mondragones, Elisa Valero Ramos, Granada, Spain, 2006

The project's most radical decision was to tie in with what was already there. The façade is shortened to make room for lacquered GRC panels in various colors. The use of color brings a certain vibration to what was already there. High windows face the morning sun, which is filtered in summer by the leaves of trees on the riverbank; and facing west, a glass wall protected by a concrete marquee opens onto a white garden.

材料：玻璃纤维增强水泥喷漆板

Material: lacquered GRC panels

芬芳的花草地
A FRAGRANT LAND OF GRASS AND FLOWERS

 栗色 Maroon

 猩红 Crimson

 橙色 Orange

 暗绿色 Dark Green

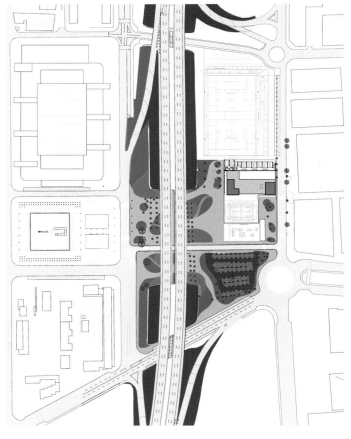

总平面图 SITE PLAN

© Pedro Pegenaute

© Pedro Pegenaute

马丁特小学，Mestura Arquitectes，西班牙，巴塞罗那，科内拉，2010年

在通往小学教室走廊的南侧是一面由瓷砖构成的幕墙，它为建筑提供了被动式遮阳。彩砖的大小300 mm x 200 mm或100 mm x 300 mm，厚22 mm，它们以互相呼应的角度拼接在一个垂直的平面上（向外或向内突出10 cm）。阳光照射最多的部分闪耀着两种色系的三种颜色。朝西的部分呈现了三种春天般的绿色，而朝东的部分则是由三种质朴的秋天般的色彩组成。

Martinet Primary School, Mestura Arquitectes, Cornellà, Barcelona, Spain, 2010

A screen has been created out of stoneware ceramic tiles to afford passive protection from the sun on the south side of the corridors that lead to the primary classrooms. The tiles are 300 mm x 200 mm/100 mm x 300 mm and 22 mm thick and are set at right angles to each other in a vertical plane (protruding outwards or inwards by 10cm). The sides that are most exposed to the sun have been glazed in two ranges of 3 different colors. The west-facing sides are in a range of 3 "spring" greens whilst the east-facing sides are in a combination of 3 earthy "autumn" colors.

材料：陶瓷、混凝土、不锈钢、石灰砂浆、增塑泡腾添加剂

Materials: ceramics, concrete, stainless steel, lime mortar, plasticized aeration additive

© Pedro Pegenaute

© Pedro Pegenaute

暗绿色 Dark Green

 橄榄褐色 Olive Drab

 绿黄色 Green Yellow

 猩红 Crimson

马丁特小学，Mestura Arquitectes，西班牙，巴塞罗那，科内拉，2010年

瓷砖的各个侧面均有凹槽，纵横交接处通过直径为6 mm 的不锈钢环加固。框架间留有7个纵向伸缩接缝，中间设有水平支撑。另外还有3个与幕墙同高的垂直支撑和两端的侧面支撑。这些接缝2 cm 厚，由石灰砂浆构成，并且加入了经过半干处理的增塑泡腾添加剂。

Martinet Primary School, Mestura Arquitectes, Cornellà, Barcelona, Spain, 2010

The tiles have a groove all around their sides that means the vertical and horizontal joints were able to be strengthened by means of 6mm diameter stainless steel rings. A total of 7 vertical expansion joints were created, with horizontal bracing in the middle framework, 3 vertical bracings through the full height of the screen and side bracings at the two ends. The joints are 2 cm thick and have been made in lime mortar with a plasticized aeration additive applied semi-dry.

材料：陶瓷、混凝土、不锈钢、石灰砂浆、增塑泡腾添加剂

Materials: ceramics, concrete, stainless steel, lime mortar, plasticized aeration additive

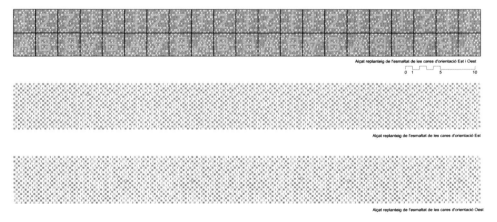

立面详图 FAÇADE DETAIL

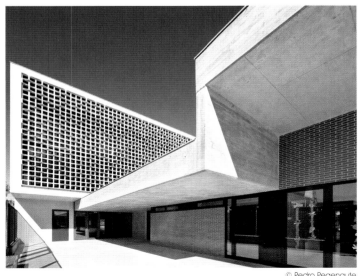
© Pedro Pegenaute

© Pedro Pegenaute

银灰色 Silver

 黄土赭色 Sienna

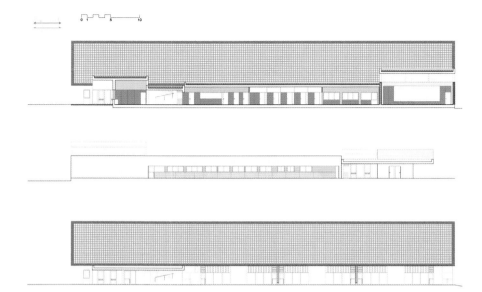

剖面图 SECTIONS

© Pedro Pegenaute

© Pedro Pegenaute

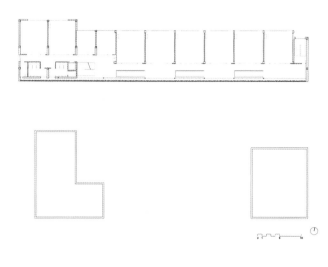

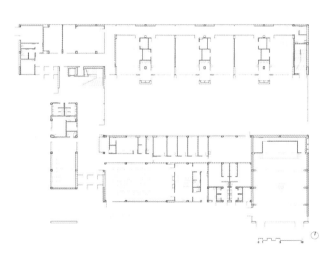

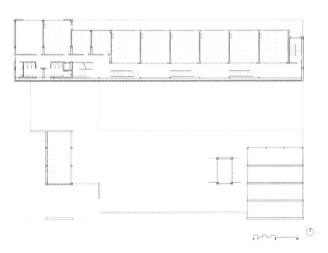

平面图 PLANS

© Pedro Pegenaute

马丁特小学,Mestura Arquitectes,西班牙,巴塞罗那,科内拉,2010年

大量玻璃窗的应用,使人们在建筑内部也能看到色彩斑斓的瓷砖墙。室内墙面用黑白两色粉刷,再无需做其他任何装饰。内部的双层立面控制了光线的照射,阳光穿过瓷砖所形成的空隙照进室内,光影效果随时间而变化。

Martinet Primary School, Mestura Arquitectes, Cornellà, Barcelona, Spain, 2010

A large number of glass windows enable people to see the colorful ceramic wall from inside. The interior walls are therefore painted in black and white, and no further decoration is needed. The double-façade in the interior serves to control the light. The sunshine leaks into the interior through the gaps formed by the ceramic tiles, creating a play of light and shadow changing over time.

材料:陶瓷、混凝土、不锈钢、石灰砂浆、增塑泡腾添加剂、涂料、玻璃、木材

Materials: ceramics, concrete, stainless steel, lime mortar, increasing plastic effervescent additives, coating, glass, wood

童年的世界
THE WORLD OF CHILDHOOD

 粉红 Pink

 猩红 Crimson

 深橙色 Dark Orange

 黄绿色 Yellow Green

 淡绿色 Light Green

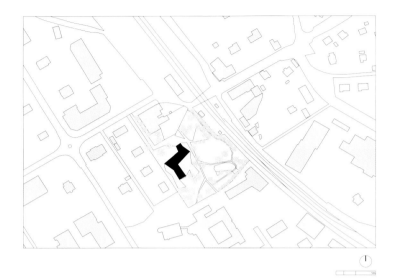

总平面图 SITE PLAN

© Hannes Henz

© Hannes Henz

蒙泰幼儿园，Bonnard Woeffray Architectes，瑞士，蒙泰，2008年

　　蒙泰幼儿园彩色墙面撑起的是稍显倾斜的屋顶。虽然是传统的建筑，外部覆层的灵感来自于童年的世界。外墙运用玩具般的色彩相互搭配，木材板条包括粉红色，橙色，红色和绿色。

Monthey Kindergarten, Bonnard Woeffray Architectes, Monthey, Switzerland, 2008

Monthey Kindergarten has a gently slanting roof which is supported by the colorful walls. While the construction is traditional, the exterior cladding was inspired by the world of childhood. The façades consist of timber slats finished in an array of toy-like colors, including pink, orange, red and green.

材料：木板、涂料

Materials: timber slats, paint

闪兰色 Dodger Blue

 紫色 Purple

 暗海洋绿 Dark Sea Green

© Hannes Henz

© Hannes Henz

黄褐色 Khaki

 闪光深绿 Lime Green

 暗海洋绿 Dark Sea Green

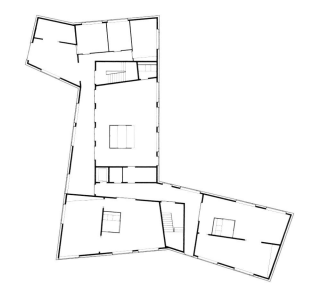

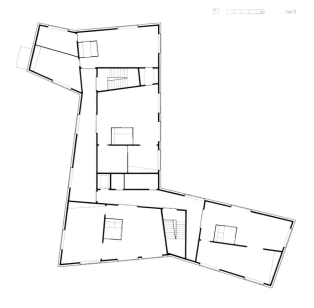

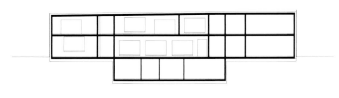

平面图 PLANS

剖面图 SECTION

© Hannes Henz

© Hannes Henz

蒙泰幼儿园，Bonnard Woeffray Architectes，瑞士，蒙泰，2008年

室内空间被分为6个独立的单元，每个单元都使用一种颜色和主题。形式各异的窗子点缀其中，大小也各不相同，各个年龄段的孩子都能找到自己的视角。室内充满了欢乐又俏皮的氛围。彩色地板和顶棚富有韵律层次，与多种颜色交相呼应。

Monthey Kindergarten, Bonnard Woeffray Architectes, Monthey, Switzerland, 2008

The inner space is divided with six separate units, each with a distinct color and theme. The variety in the positions of the windows, all identical in size, contributes to a range of perceptions and facilitates visual contact for all ages. An happy and playful ambience is created in the house. The interior is composed of a rhythmic succession of colored floors and ceilings, with as many colors as there are units.

材料：木板、涂料、瓷砖、玻璃

Materials: planks, coating, ceramic tiles, glass

风中的芦苇塘
THE REED FIELD IN THE WIND

 钢蓝 Steel Blue

 猩红 Crimson

 金色 Gold

 茶色 Tan

 粉红 Pink

 亮菊黄 Light Goldenrod Yellow

维的亚兰卡尔理工学院学习中心X大厦 二期工程，Planet 3 Studios Architecture，印度，孟买，2009年

设计师的灵感来自"芦苇塘"——"莲花"，层层粉中透白的"花瓣"包裹着五间弧形的实验室；黑色、灰色的"枝杈"将"莲花"衬托得栩栩如生；由淡黄色、褐色的六边形相间而成的屋顶和墙面装饰，不禁使人联想到了蜂巢的形态；一条巨大的波浪形曲线贯穿二楼始终，围合其中的是不规则分布的教室；灰色、原木色、大红色、朱红色彩条间隔的木墙和红白两色构成的曲线吊顶辉映成趣，似是倒映着荷花的水面，荡漾开圈圈涟漪；此外，还有几何形装饰的树木、色彩鲜明的坐凳、层次清晰的"绿叶"……所有的"野趣"都有待于你的发现和思考。

'X' Block Vidyalankar Annex – Phase II, Planet 3 Studios Architecture, Mumbai, India, 2009

The architects gained their inspiration from the "reed field". The interior is dominated by a gigantic lotus flower with layers of pink petals with a touch of white, cladding over five arc-shaped laboratories; black and gray "braches" serve as a foil for the vivid lotus flower; the ceiling and wall are decorated with faint yellow and brown hexagons which are reminiscent of a honeycomb; the irregular classrooms are enclosed by a giant undulating curving wall that stretches over the second floor; the wooden wall embellished with gray, burly wood, scarlet and vermilion stripes and the red-white suspended ceiling form a delightful contrast with each other, just like rippling water surface reflecting the lotus flowers; the trees decorated with geometries, brightly-colored stools, clearly-layered green leaves…all the "rustic charms" are waiting for you to discover and ponder.

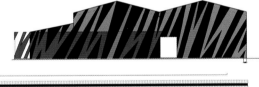

立面图 ELEVATIONS

材料：涂料、木板、金属板材、玻璃

Materials: paint, planks, metal panels, glass

© Mrigank Sharma, India Sutra

© Mrigank Sharma, India Sutra

© Mrigank Sharma, India Sutra

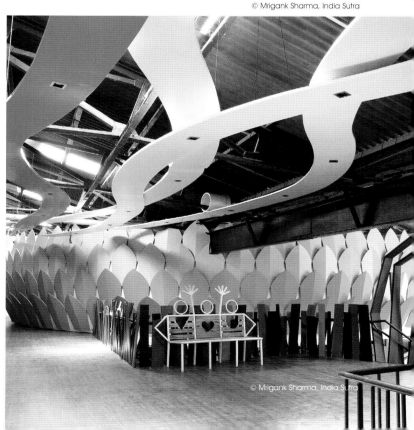

© Mrigank Sharma, India Sutra

青蛙王子
FROG KING

 浅黄色 Light Yellow

 金色 Gold

 茶色 Tan

 粉红 Pink

 淡紫色 Lavender

 暗海洋绿 Dark Sea Green

 印度红 Indian Red

 闪光深绿 Lime Green

 闪兰色 Dodger Blue

深红色 Dark Red

暗蓝色 Dark Blue

"青蛙王子"，Splitterwerk.Grazt Architektur ZT GmbH，奥地利，施泰尔马克州，格拉茨市，2010年

建筑的表面被不同的颜色所覆盖：白色、红色、黄色、绿色等，细细数来一共有26种颜色。其覆盖是彻底性的，无论是建筑立面、屋顶还是山墙，均被披上了五颜六色的"彩装"。每一块镶板15 cm高、60 cm长，这些抽象的面板通过不断地重复来扩大其面积。它们环绕着建筑体，营造出一种不断扩散的色彩效果。同样，由于彩色水泥镶板的覆盖，建筑表面那些破碎的结构统统被遮挡起来。远远看去，建筑像是小孩子用手中摆弄的彩色积木搭建而成，甚至朝向室外的百叶窗也被设计成了彩色的条纹图案。当用户将百叶窗拉下时，外面的人甚至无法确认窗户的存在，因为它已经完全被淹没在周围的彩色镶板之中。设计师就是用这种方法，来实现"青蛙"变成"王子"的。

Frog King, Splitterwerk.Grazt Architektur ZT GmbH, Graz, Steiermak, Austria, 2010

The surface of the building is covered with fiber cement panels in various colors, such as white, red, yellow, green and so forth, 26 colors in all. From the façades, roof to the pediment, the "colorful coat" covers every plane. These abstract panels, measuring 15 x 60 cm each, are alternatively applied to enclose the entire building, creating an effect of dispersion. By the same token, the cracked structures on the surface are all sheltered. From afar, the building seems to be comprised of colored building blocks that kids always play. Even the outward louvers are designed with colored stripe patterns. When they are shut, people can not even recognize its existence, because they are completely submerged in the surrounding colorful panels. This is the right way in which the designer turns the "frog" into a "king".

材料：15 cm×60 cm纤维水泥板

Material: fiber cement board

© Nikolaos Zachariadis

© Nikolaos Zachariadis

草绿色 Lawn Green

 猩红 Crimson

 矢车菊蓝 Cornflower Blue

© Nikolaos Zachariadis

© Nikolaos Zachariadis

橙色 Orange

 深橙色 Dark Orange

 闪光深绿 Lime Green

© Nikolaos Zachariadis

© Nikolaos Zachariadis

"青蛙王子",Splitterwerk.Grazt Architektur ZT GmbH,奥地利,施泰尔马克州,格拉茨市,2010年

设计师在不同功能区的墙面上刷上了不同的颜色,比如办公室的暗黄色、走廊的红色、研讨室的蓝色、休息室的绿色……这一方面舍弃了抽象而枯燥的文字导向系统,另一方面也为用户提供了更加明确的视觉引导。当然,更重要的是,色彩的使用让室内空间变得焕然一新,之前死气沉沉的感觉被清新活泼的气氛所代替。当用户推开窗户,这样的感觉会愈加明显,明亮的阳光与墙面上鲜艳的色彩水乳交融,产生出深浅不一、丰富的光色变化。这样的视觉感受不由得让人想起莫奈的那些非凡的印象派作品,而亦真亦幻之感又似乎带有了卢梭作品中的超现实主义气质。

Frog King, Splitterwerk.Grazt Architektur ZT GmbH, Graz, Steiermak, Austria, 2010

The designer distinguishes each functional zone by different colors: dark yellow for the office, red for the corridor, blue for the seminar room, green for the common room...In this way the abstract, boring literal guiding system is replaced by a more explicit visual guiding system. More importantly, the interior space takes on a new look with the bright colors while the previous spiritless feeling is superseded by a clear, lively atmosphere. The feelings become even clearer when the client opens the windows. The bright sunshine and the vivid colors are as well blended as milk and water, creating rich variations. It reminds people of the prominent works by Monet while its real yet imaginary sense seems to contain the quality of super-realism works by Rousseau.

材料:涂料、百叶窗、玻璃

Materials: paint, blinds, glass

彩色条纹的欢快韵律
THE GAILY RHYTHM OF COLORFUL STRIPES

 深红色 Dark Red

 深粉红 Deep Pink

 亮粉红 Light Pink

 珊瑚 Coral

 亮天蓝色 Light Sky Blue

 灰菊黄 Pale Goldenrod

总平面图 SITE PLAN

© José Hevia

© José Hevia

比尼萨莱姆学校综合建筑，RipollTizon，西班牙，马略卡岛，2011年

底层立面的用色意图是为对面操场上的孩子营造背景。孩子们缤纷的服装和充满朝气的身影将与立面上的夺目的色块融为一体。立面的色彩图案源自于附近学校衣着艳丽的学生在操场玩耍的生动照片。这些色彩和在场景中所占的比例构成了该项目的色彩图案。不同宽度的全高式纵向彩条按照色彩的韵律配列。最终，生机勃勃、变幻多彩的立面遵循了孩子们的设计逻辑。

Binissalem School Complex, RipollTizon, Mallorca, Spain, 2011

The intention of using color in the ground floor façade facing the playground areas is to create a background for the children. Their vibrant clothing and movement will blend with the bold blocks of color of the façade. The color pattern of the façade is inspired by photographies of students wearing colorful clothing on the playground of some schools nearby. The color palette and density in these scenes are used to develop the color patterns of this project. It consists of full height vertical stripes that are arranged following designed rhythms of color overlapping a base of stripes in different widths. In the end, the vibrant and varied image of the façade responds to a design logic inspired in the children.

材料：有机无污染涂料、水泥砂浆

Materials: organic non-polluting paints over a three layer rendering with cement mortar

© José Hevia

© José Hevia

彩虹雨
RAINBOW SHOWER

 金色 Gold

 纯黄 Yellow

 黄绿色 Yellow Green

 粉红 Pink

 弱绿宝石 Pale Turquoise

 闪兰色 Dodger Blue

 暗紫罗兰 Dark Violet

 猩红 Crimson

巢鸭信用银行江古田支行,Emmanuelle Moureaux Architecture + Design,日本,东京,2012年

建筑场地位于商店林立的商业街区,建筑退后界限2m,外围以木材铺装,并排立了许多9m高的彩杆。这29根杆子倒映在建筑的玻璃里面上,与室内随意排列的19根杆子自然地相呼应。这一如彩虹雨般美妙的风景为城市增添丰富的色彩和趣味十足的空间。该地与城市生活密切相关,车水马龙,人行道狭窄,这激发了建筑师的灵感,通过建筑室内外空间的融合来表达这一紧密联系。

Sugamo Shinkin Bank / Ekoda Branch, Emmanuelle Moureaux Architecture + Design, Tokyo, Japan, 2012

The site is located in a commercial district with many stores. The building is offset approximately 2 meters from the property line, and the timber-decked peripheral space is filled with colorful 9 meter-tall sticks. These 29 exterior sticks, reflected on the transparent glazed façade, mix naturally with the 19 interior sticks placed randomly inside the building. This rainbow shower returns colors and some room for playfulness back to the town. The site's closeness to the town's activities – also the heavy traffic and narrow sidewalk – inspired the architect to express this proximity in the building by merging the exterior and interior.

材料:外饰面 铝板涂氟树脂漆、挤塑成型水泥预制板涂覆氟树脂漆
　　　室外地面 甲板
　　　彩杆 不锈钢涂氟树脂漆

Materials: Exterior Finish: aluminum plate fluoro-resin paint finish, extruded cement panel fluoro-resin paint finish
External Floor: deck
Sticks: structural SUS, Fluoro-resin paint finish

© Daisuke Shima / Nacasa & Partners Inc.

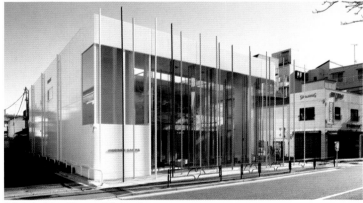
© Daisuke Shima / Nacasa & Partners Inc.

© Daisuke Shima / Nacasa & Partners Inc.

橙色 Orange

 黄褐色 Khaki

 黄绿色 Yellow Green

 浅粉红 Light Pink

 天蓝色 Sky Blue

 暗紫罗兰 Dark Violet

 印度红 Indian Red

© Daisuke Shima / Nacasa & Partners Inc.

© Daisuke Shima / Nacasa & Partners Inc.

巢鸭信用银行江古田支行，Emmanuelle Moureaux Architecture + Design，日本，东京，2012年

进入建筑，游客会感到他们仍然身处一个通往银行内部的室外庭院之中。在这里，室内和室外仍然融为一体。走过笼罩在玻璃墙中的庭院进入室内，一个自然光线充沛，像咖啡厅一样的开放空间呈现在眼前。栽种在庭院中的竹子耸向天空，与彩杆珠联璧合。室外的露台、室内的开放空间、外部的庭院以及室内的出纳柜台构成了4个层次的空间。这些层次倒映在玻璃幕墙上，与复杂的光影相映成趣，增添了空间的进深感。

Sugamo Shinkin Bank / Ekoda Branch, Emmanuelle Moureaux Architecture + Design, Nerima-ku, Tokyo, Japan, 2012

Entering the building, the visitors would notice that they are still in an exterior courtyard leading to the bank's interior. Here also, the inside and outside are integrated. Walking around the glazed courtyard inside, there is a cafe-like open space filled with natural light. The bamboos in the courtyard extend skyward in concert with the colorful sticks. The exterior deck space, interior open space, exterior courtyard, and the interior teller counters compose four layers of spaces. The layers are reflected on the glazing, and, combined with complex shadows, creating depth in the space.

材料：室内地面 地毯、PVC地板
       墙壁 PVC薄膜+彩色黏膜
       顶棚 AEP漆+彩色黏膜

Materials: Interior Floor: carpet, vinyl flooring
      Wall: PVC Film + adhesives colored films
      Ceiling: AEP paint + adhesives colored films

彩色方块世界
A WORLD OF COLORFUL SQUARES

 靛青 Indigo

 闪兰色 Dodger Blue

 亮天蓝色 Light Sky Blue

 闪光深绿 Lime Green

 黄绿色 Yellow Green

 猩红 Crimson

 纯黄 Yellow

 亮粉红 Light Pink

巢鸭信用银行新座支行,Emmanuelle Moureaux Architecture + Design,日本,崎玉,2009年

　　该项目旨在颠覆人们心中金融机构的印象,创造一个崭新的形象。设计的重点围绕"方块"展开,除了在设计中融入方块形状,该建筑被认为是一个供人们聚集的公共广场。这些方块的颜色起到了重要的作用:建筑立面上的公司标识由24种颜色拼成,在主街道清晰可见,成为该地区的象征。

Sugamo Shinkin Bank / Niiza Branch, Emmanuelle Moureaux Architecture + Design, Saitama, Japan, 2009

This project sought to create a whole new look that refreshes the current image of this financial institution. The key concept revolves around squares - besides incorporating square shapes, the building was conceived as a sort of public square where people gather. The colors of these squares play an important role: the logo on the façade of the building features as many as 24 colors visible from the main street, becoming a symbol for the area.

材料:钢筋混凝土、玻璃

Materials: reinforced concrete, glass

© Hidehiko Nagaishi

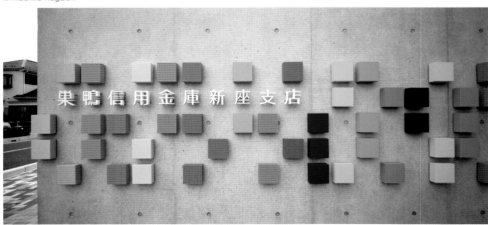
© Hidehiko Nagaishi

立面图 ELEVATION

亮粉红 Light Pink

 深粉红 Deep Pink

 猩红 Crimson

 深洋红 Dark Magenta

© Hidehiko Nagaishi

© Hidehiko Nagaishi

© Hidehiko Nagaishi

黄绿色 Chartreuse

 森林绿 Forest Green

 闪兰色 Dodger Blue

 纯蓝 Blue

巢鸭信用银行新座支行，Emmanuelle Moureaux Architecture + Design，日本，崎玉，2009年

这些色彩延伸至室内，它们不仅仅是装饰，更承担着内部空间功能区隔的作用，将大堂、会议区、自动柜员机等区域自然地分隔开来。紫色系的渐变色块区域是自动柜员机的所在，绿色系渐变色块则标示出大堂，而红黄色系渐变色块围出了会议、洽谈的小空间。

Sugamo Shinkin Bank / Niiza Branch, Emmanuelle Moureaux Architecture + Design, Saitama, Japan, 2009

These colors welcome customers as they enter the building, continuing on the inside and serving as natural dividers between lobby, meeting space, ATM and so on. The area with purple gradient color squares defines the ATM area, the area with green gradient color squares defines the lobby, the area with red and yellow gradient color squares defines the meeting space, etc.

© Hidehiko Nagaishi

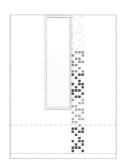

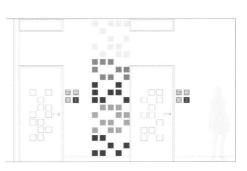

立面详图 FAÇADE DETAIL

© Hidehiko Nagaishi

七彩缤纷 | RAINBOW COLOR

彩虹千层派
RAINBOW MILLE - FEUILLE

 金色 Gold

 橙红色 Orange Red

 粉红 Pink

 绿黄色 Green Yellow

 闪光深绿 Lime Green

 深天蓝 Deep Sky Blue

北立面图 NORTH ELEVATION

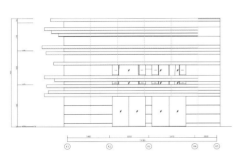

西南立面图 SOUTHWEST ELEVATION

© Daisuke Shima / Nacasa & Partners Inc.

© Daisuke Shima / Nacasa & Partners Inc.

巢鸭信用银行志村支行，Emmanuelle Moureaux Architecture + Design，日本，东京，2011年

　　巢鸭信用银行没有采用传统银行那种冰冷冷、规规整整的设计，设计师利用层叠的造型、彩虹般的色彩，将银行严肃呆板的氛围变得友好、多姿多彩、生动活泼。建筑的12层色彩仿佛构成彩虹似的千层派。一层层色彩鲜艳的悬挑板如彩虹般错列叠置，从立面上向外窥探，仿佛在欢迎着游客的到来。这些色彩反射到白色的墙面上，留下一层朦胧的彩色光晕，给人以温馨、柔和、迷人的感觉。当夜幕降临，彩板被微弱的灯光点亮，在不同的季节，一天中不同的时间都有变化，幻化出无尽的风景。

Sugamo Shinkin Bank / Shimura Branch, Emmanuelle Moureaux Architecture + Design, Tokyo, Japan, 2011

The architects of Sugamo Shinkin Bank abandoned the stiff design of traditional bank buildings. Instead, they create a friendly, colorful and lively atmosphere with colored layers and rainbow-like colors in the traditionally serious and dull bank. Twelve layers of color comprise the rainbow-like mille-feuille. The rainbow-like stack of colored layers, peeks out alluringly from the façade to welcome visitors. Reflected onto the white surface, these colors leave a faint trace over it, creating a warm, glamorous, and captivating yet gentle feeling. At night, the colored layers are faintly illuminated. The illumination varies according to the season and time of day, conjuring up myriad landscapes.

材料：外饰面 铝板涂氟树脂漆
　　　室外地面 甲板

Materials: Exterior Finish: aluminum plate Fluoro-resin paint finish
　　　　　External Floor: deck

© Daisuke Shima / Nacasa & Partners Inc.

猩红 Crimson

浅粉红 Light Pink

深橙色 Dark Orange

金色 Gold

深天蓝 Deep Sky Blue

海军蓝 Navy

© Daisuke Shima / Nacasa & Partners Inc.

巢鸭信用银行志村支行，Emmanuelle Moureaux Architecture + Design，日本，东京，2011年

建筑内部有3个椭圆形的天窗，阳光从天窗透入，室内空间沐浴在柔和的光线之中。人们走进建筑时，会不由自主地抬头仰望那片露出的天空。开放的天空及开阔的感觉会使人不由地深深呼吸，感觉神清气爽。

顶棚装饰着彩色的蒲公英绒毛的图案，看起来好像飘飞在空中一样。在欧洲，有一个古老的传统，即在吹蒲公英毛球的时候默默地许愿。在清风的吹拂下，一簇簇绒毛在空中舞蹈嬉戏，慢慢飘落。

Sugamo Shinkin Bank / Shimura Branch, Emmanuelle Moureaux Architecture + Design, Tokyo, Japan, 2011

Upon entering the building, three elliptical skylights bathe the interior in a soft light. Visitors spontaneously look up to see a cut-out piece of the sky that invites them to gaze languidly at it. The open sky and sensation of openness prompts you to take deep breaths, refreshing your body from within.

The ceiling is adorned with colorful dandelion puff motifs that seem to float and drift through the air. In Europe, there is a long and cherished custom of blowing on one of these fuzzy balls while secretly making a wish. Bits of fluffy down gently dance and frolic in the air, carried by the wind.

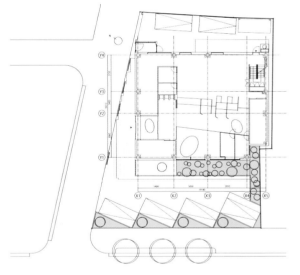

一层平面图 FIRST FLOOR PLAN

材料：室内地面 地毯、PVC地板
　　　墙壁 PVC薄膜+彩色黏膜
　　　顶棚 AEP漆+彩色黏膜

Materials: Interior Floor: carpet, vinyl flooring
　　　Wall: PVC Film + adhesives colored films
　　　Ceiling: AEP paint + adhesives colored films

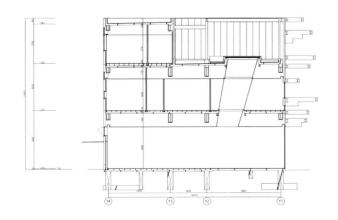

剖面图 SECTION

魔幻树叶
MAGIC LEAVES

 橙色 Orange

 火砖色 Fire Brick

 闪光深绿 Lime Green

 闪兰色 Dodger Blue

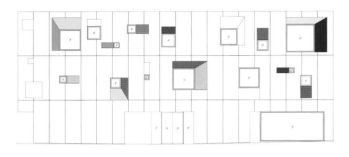

立面图 ELEVATION

© Daisuke Shima / Nacasa & Partners Inc.

© Daisuke Shima / Nacasa & Partners Inc.

巢鸭信用银行常盘台支行,Emmanuelle Moureaux Architecture + Design,日本,东京,2010年

建筑师以树叶主题为基础,试图创造出清新的空间,让顾客有一种自然、充满活力的感觉。建筑立面上有很多树木轮廓,还设置了大大小小14种颜色的窗户,排列方式独特,极具韵律感,将立面本身转变成了一处标志。

Sugamo Shinkin Bank / Tokiwadai Branch, Emmanuelle Moureaux Architecture + Design, Tokyo, Japan, 2010

By basing the design around leaf motifs, the architects sought to create a refreshing space that would welcome customers with a natural, rejuvenated feeling. The façade of the building features silhouettes of trees and an assortment of both large and small windows in 14 different colors arranged in a distinctive, rhythmical pattern that transforms the façade itself into signage.

材料:外饰面 铝板涂氟树脂漆
室外地面 甲板

**Materials: Exterior Finish: aluminum plate Fluoro-resin paint finish
External Floor: deck**

© Daisuke Shima / Nacasa & Partners Inc.

猩红 Crimson

 亮粉红 Light Pink

 深橙色 Dark Orange

 金色 Gold

 闪兰色 Dodger Blue

 海军蓝 Navy

巢鸭信用银行常盘台支行，Emmanuelle Moureaux Architecture + Design，日本，东京，2010年

一层的开放空间摆放着14种不同颜色椅子，令人耳目一新。"生长"在墙体和玻璃窗上面白色树枝上的24种不同颜色的树叶，与庭院内真正的树木上的天然叶子相互重叠，给人一种正身处魔幻森林中的感觉。

Sugamo Shinkin Bank / Tokiwadai Branch, Emmanuelle Moureaux Architecture + Design, Tokyo, Japan, 2010

On the first floor, there is an open space laid out with chairs in 14 different colors, giving people a pleasant feeling. A constellation of leaves in 24 different colors growing on the white branches of the walls and glass windows overlaps with the natural foliage of the real trees in the courtyards, creating the sensation of being in a magical forest.

材料：室内地面 地毯、PVC地板
　　　墙壁 AEP漆+彩色黏膜
　　　顶棚 AEP漆

Materials: Interior Floor: carpet, vinyl flooring
　　　　　Wall: AEP paint + adhesives colored films
　　　　　Ceiling: AEP paint

© Daisuke Shima / Nacasa & Partners Inc.

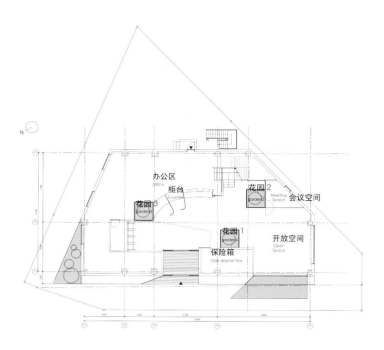

一层平面图 FIRST FLOOR PLAN

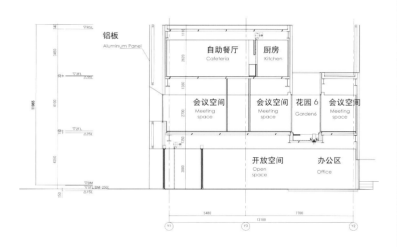

剖面图 SECTION

色彩的王国
A KINGDOM OF COLORS

 橙色 Orange

 珊瑚 Coral

 橙红色 Orange Red

 粉红 Pink

 深红色 Dark Red

 绿黄色 Green Yellow

 闪兰色 Dodger Blue

 暗兰花紫 Dark Orchid

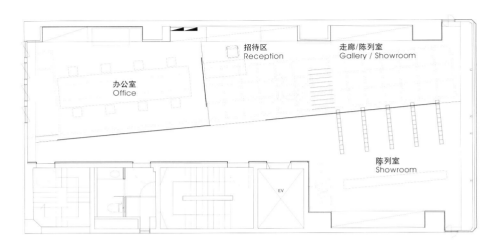

平面图 PLAN

© Hidehiko Nagaishi

© Hidehiko Nagaishi

中川化工CS设计中心，Emmanuelle Moureaux Architecture + Design，日本，东京，2007年

当展品色彩丰富之极的时候，展厅、展台最好就用虚无的白色了。中川化学CS设计中心功能作为一个互动空间，人们可以在那里尝试一下多达1200个不同颜色的薄膜样品，就像漫步在色彩的王国。

Nakagawa Chemical CS Design Center, Emmanuelle Moureaux Architecture + Design, Tokyo, Japan, 2007

When the colors of the exhibits are extremely rich, it is best to adopt the color of nihility—white. The Nakagawa Chemical CS Design Center functions as an interactive space where people can try out as many as 1,200 different colored film samples just like browsing through a kingdom of colors.

材料： 地面 橡木地板、黑色地毯
墙壁 白色丙烯酸树脂漆
顶棚 白色丙烯酸树脂漆

Material: Floor: oak flooring, black tile carpet
Wall: white acrylic enamel paint
Ceiling: white acrylic enamel paint

© Hidehiko Nagaishi

© Hidehiko Nagaishi

© Hidehiko Nagaishi

幻彩殿堂
COLORFUL PALACE

 银灰色 Silver

 浅黄色 Light Yellow

 橙色 Orange

 暗金菊黄 Dark Goldenrod

 粉红 Pink

 中紫色 Medium Purple

 纯红 Red

贝里约扎尔幼儿园，Javier Larraz & Iñigo Beguiristain & Iñaki Bergera，西班牙，纳瓦拉，贝里约扎尔，伊洛塔街，2012年

建筑周遭的彩色百叶窗灵感来自于孩子们画画用的蜡笔，能够起到调节教室内光线的功效，同时屏蔽了来自建筑末端操场的噪声与干扰。孩子们就像花朵，当然需要充足的阳光，于是建筑屋顶高高隆起的天窗为室内带来大量光照，同时也赋予了建筑的体形变化，增添了趣味性。

Nursery School in Berriozar, Javier Larraz & Iñigo Beguiristain & Iñaki Bergera, Calle Errota, Berriozar, Navarra, Spain, 2012

The brightly colored louvers on each elevation were inspired by children's crayons used for drawing. They can moderate light into the classroom whilst forming a protective screen around the two playgrounds located at the ends of the building. The children are like flowers which need abundant sunshine. Accordingly, the generous overhead skylight brings plenty of natural light into the interior. It also endows the building a change in form and a playful character.

材料：混凝土、铝、木材

Materials: concrete, aluminum, wood

© Iñaki Bergera

© Iñaki Bergera

© Iñaki Bergera

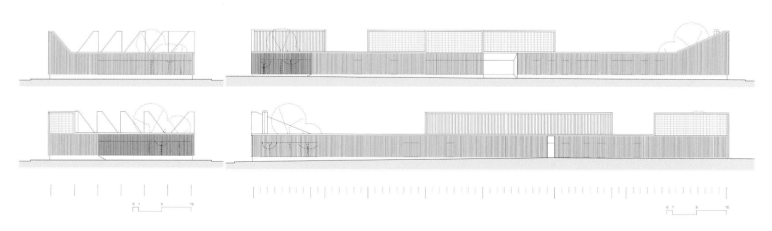

立面图 ELEVATIONS

南十字星闪耀
THE SHINNING SOUTHERN CROSS

 黛青色 Dark Cyan

 绿黄色 Green Yellow

 深洋红 Dark Magenta

 深橙色 Dark Orange

 橙色 Orange

 亮天蓝色 Light Sky Blue

 深灰色 Dark Gray

南十字星圣母学校的图书馆与大厅，Fulton Trotter Architects，澳大利亚，昆士兰州，2011年

新入口处的遮阳棚采用了与原大厅屋顶相同的钢桁架，实现了新老建筑在视觉的一致性。下面的支撑柱显著地呈现了学校的标志"南十字星"的造型。外墙的元素如排摆的书本一般，在色彩上营造出了有趣的视觉效果。室内装饰采用了波纹状的顶棚和钢结构，体现了与多尔比地区相呼应的工业美学。缤纷的色彩彰显出孩子们青春的活力，营造出一个轻松活泼的氛围。

Our Lady of the Southern Cross College – Library & Hall, Fulton Trotter Architects, Queensland, Australia, 2011

Expressed steel truss roof framing used on the existing hall was continued for the new entry awning and library, which visually ties together existing and new work. Supporting columns implore a strong visual element of the College symbol – "Southern Cross". Visual games were played, with wall elements arranged like stacked books. The theme of bright colors found in the existing primary school was used and continued. An industrial aesthetic relating to the Dalby region was developed within finishes of corrugated steel ceilings and expressed steel structure. The varied colors manifest the youthful vigor of the children, thus creating a relaxing, vivid ambience.

材料：钢材、玻璃

Materials: steel, glass

平面图 图书馆&门厅 PLAN - LIBRARY & HALL

© Carmichael Builders

© Carmichael Builders

© Carmichael Builders

© Carmichael Builders

自然之园
THE KINDERGARTEN OF NATURE

 粉红 Pink

 茶色 Tan

 黄土赭色 Sienna

卡塔丽娜富兰克潘幼儿园，Randić & Turato，克罗地亚，克尔克岛，2009年

淡淡的粉红色铺满墙面，宛若花朵般娇嫩，与石头的粗糙感形成鲜明的对比。坚硬的石墙仿佛就是为了保护那一抹粉红而存在。幼儿园从一开始，就打算跟周围的环境"划清界限"。一圈粗糙的石墙将幼儿园团团围起，形成一个幼儿园的封闭小王国，在内部布置出单元房及其周围开放式的花园。

Kindergarten Katarina Frankopan, Randić & Turato, Krk Island, Croatia, 2009

The walls are painted pale pink, gender and lovely like the petals. They are in sharp contrast with the rough stone walls, which seem to exist only to protect the pink color. In such unattractive neighborhood, this kindergarten is shaped as enclosed and introverted insula, surrounded with soaring stone walls. Inside of this small town-kindergarten, units-houses are combined with open gardens, placed next to pedestrian communications.

材料：石头、涂料、玻璃

Materials: stone, paint, glass

© Sandro Lendler

© Sandro Lendler

© Sandro Lendler

© Sandro Lendler

浅灰色 Light Gray

 深粉红 Deep Pink

 金色 Gold

 暗灰蓝色 Dark Slate Blue

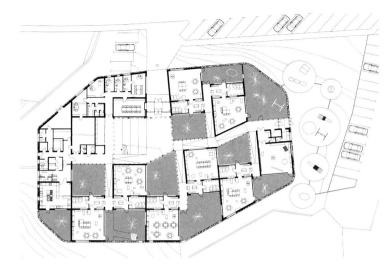

一层平面图 FIRST FLOOR PLAN

剖面图 SECTION

© Sandro Lendler

© Sandro Lendler

卡塔丽娜富兰克潘幼儿园，Randić & Turato，克罗地亚，克尔克岛，2009年

室内运用了红、黄、紫等大胆且轻松的颜色。幼儿期的儿童好奇心强，喜欢鲜艳的色彩，在涂涂画画中，认识色彩对幼儿是一种不可抵挡的诱惑。有意识地培养他们的色彩感知能力，可以引起幼儿愉快的体验，获得美的享受，也有利于幼儿良好性格的形成和智力的发展。

Kindergarten Katarina Frankopan Island, Randić & Turato, Krk Island, Croatia, 2009

Children are curious in infancy and are fond of bright colors. It is an irresistible temptation for children to recognize colors through painting. Consciously developing children's ability to perceive colors will bring them a pleasant experience and the enjoyment of beauty. Meanwhile, it also helps them to build good character and develop their intelligence.

材料：玻璃、木材、钢材、涂料

Materials: glass, wood, steel, paint

彩虹屋
RAINBOW HOUSE

 火砖色 Fire Brick

 深橙色 Dark Orange

 橙色 Orange

 纯黄 Yellow

 绿黄色 Green Yellow

 森林绿 Forest Green

 深天蓝 Deep Sky Blue

 钢蓝 Steel Blue

棕榈树丛中的幼儿园，Cor & Asociados. Miguel Rodenas + Jesús Olivares，西班牙，穆西亚省，2012年

幼儿园坐落在3条道路交叉处的一块三角形场地，四周环绕着一片片零落的绿地。幼儿园主体为白色，以彩色穿孔板装饰，就像是雨后森林里的七色彩虹，是幼儿眼睛里的缤纷世界。

Kindergarten Between Palms in Los Alcazares, Cor & Asociados. Miguel Rodenas + Jesús Olivares, Murcia, Spain, 2012

Located between three roads, triangular plant, the kindergarten is surrounded by a disjointed sequence of wooded green space. The main body of the kindergarten is in white, adorned with colorful perforated panels, like a beautiful rainbow in the forest. The kindergarten is a colorful world for children.

材料：玻璃、金属

Materials: glass, metal

© David Frutos

© David Frutos

天蓝色 Sky Blue

 金菊黄 Goldenrod

 巧克力色 Chocolate

© David Frutos

© David Frutos

橄榄褐色 OliveDrab

 金菊黄 Goldenrod

 暗海洋绿 Dark Sea Green

洛斯阿尔卡萨雷斯幼儿园，Cor & Asociados. Miguel Rodenas + Jesús Olivares，西班牙，穆西亚省，2012年

幼儿园内部墙壁为柔和的白色，装饰有彩色的卡通插画，十分可爱。地板颜色根据区域划分而不同，粉红色、粉蓝色、粉绿色的色彩十分柔和，创造出一种明快、温馨的氛围，适合幼儿学习和成长。

Kindergarten in Los Alcazares, Cor & Asociados. Miguel Rodenas + Jesús Olivares, Murcia, Spain, 2012

The internal wall of the kindergarten is painted white, adorned with very cute cartoon characters in different colors. The colors of the floors are different in different area. The colors of pink, light blue, pastel green are very gentle, creating a lively and warm atmosphere, suitable for children to learn and grow.

材料：墙壁 定向刨花板
　　　地面 亚麻地板
　　　卫生间 瓷砖
　　　Pladur涂料

Materials: Wall: OSB panel
　　　　　　Floor: Lineoleum
　　　　　　Bathroom: Cermics
　　　　　　Pladur

© David Frutos

七彩缤纷 | RAINBOW COLOR | 87

流动的幻彩韵律
THE FLOWING RHYTHMS

爱丽丝蓝 Alice Blue

粉蓝色 Powder Blue

亮菊黄 Light Goldenrod Yellow

橙色 Orange

纯红 Red

紫色 Purple

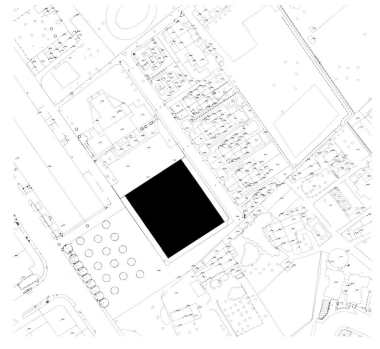

总平面图 SITE PLAN

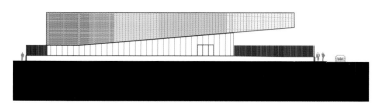

立面图 ELEVATION

Sa Indiotería 体育中心，Jordi Herrero + Sebastián Escanellas，西班牙，帕尔马，2007年

Sa Indiotería 体育中心拥有彩虹般的外立面，色彩绚丽，富有流动的韵律感。板材色彩缤纷，粉色俏皮可爱、紫色典雅高贵、橙色温暖明媚，绿色清新自然，如春光般缤纷多彩，彰显青春与活力。建筑上方是大型半透明屋顶，外壳由50 cm 宽，4 cm 厚的聚碳酸酯板构成，能够使光线自由地在室内外流动。建筑拐角处的设计也是别出心裁，摒弃了棱角分明的铝型材，采用了弯折的聚碳酸酯板。建筑下方外壳则由防阳光、低辐射率的 8+8 夹层玻璃构成。东南与西南立面的巨大悬臂框架起到了遮阳的效果。

Sports Center Sa Indiotería, Jordi Herrero + Sebastián Escanellas, Palma de Mallorca, Spain, 2007

Sports Center Sa Indiotería has a rainbow-like façade, and the gorgeous colors create a sense of flowing rhythms. The sheets incorporate different colors, the lovely pink, elegant purple, warm orange, fresh green are as bright as the spring scenery, manifesting vigor and vitality. The volume enclosure is made of polycarbonate sheets 50 cm wide and 4 cm thick. A big translucent roof made of colored polycarbonate levitates over the court allowing the light moving inside and outside freely. Special care was given to the corners. Instead of using edge profiles of aluminum, it used bent polycarbonate sheets. The lower enclosure consists of 8+8 laminated glass with solar control and low emissivity. The sun protection is guaranteed by the big size overhangs in the southeast and in the southwest façades.

材料：塑料

Material: plastic

© Jaime Sicilia / arquipress

© Jaime Sicilia / arquipress

 雪白色 Snow
 烟灰色 Smoke Gray
 茶色 Tan
 猩红 Crimson
 橙红色 Orange Red
 纯黄 Yellow
闪兰色 Dodger Blue

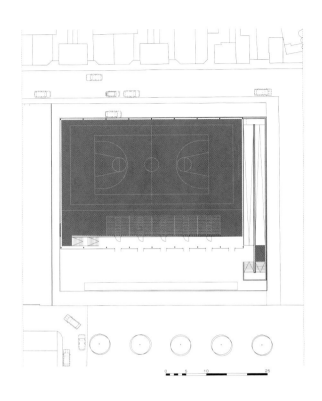

一层平面图 GROUND FLOOR PLAN

© Jaime Sicilia / arquipress

© Jaime Sicilia / arquipress

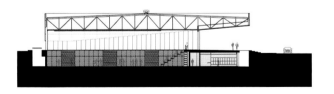
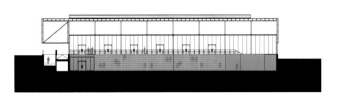

剖面图 SECTIONS

Sa Indiotería 体育中心，Jordi Herrero + Sebastián Escanellas，西班牙，帕尔马，2007年

该项目室内的色彩设计摒弃了全封闭的设计，而是通过室外色彩向室内的延伸，而将内外空间连为一体。设计师将庭院与更衣室设于地下，这种设计能够使人们从公园一路通向带顶平台、看台，一直到最下方的庭院，视野通透，一览无余。建筑未采用屋顶采光，也不是间接从北面或对角线方向采光。以上采光方式都在某种程度上将光线"锁定"了，而这里的光线是流动的，透过缤纷的外立面照射到室内的走廊、大厅、篮球馆……令原本单一的色调立刻欢快了起来。使体育中心无论在外观还是室内，都充满了勃勃生机。

Sports Center Sa Indiotería, Jordi Herrero + Sebastián Escanellas, Palma de Mallorca, Spain, 2007

Instead of the image of a hermetic pavilion, here the intention is to give continuity from "Construction" the inside space to the outside one. The building is implemented to bury the court and the locker rooms. The schema allows going through, without visual barriers, from the park to the covered platform, to the stands, and finally descending to the court. The kind of light of the sports hall is not of the type of zenith, neither indirect to the north, nor diagonal; all of them corresponding to some kind of locked light. But in here the light is the kind of flowing light which floods into the corridor, lobby, basketball gym…The previously single-colored space is activated and vigor fills in both interior and exterior.

材料：塑料、混凝土、玻璃

Materials: plastic, concrete, glass

© Jaime Sicilia / arquipress

© Jaime Sicilia / arquipress

© Jaime Sicilia / arquipress

色彩斑斓的住宅
COLORFUL HOUSING

 森林绿 Forest Green

 纯黄 Yellow

 深橙色 Dark Orange

 猩红 Crimson

 暗蓝色 Dark Blue

总平面图 SITE PLAN

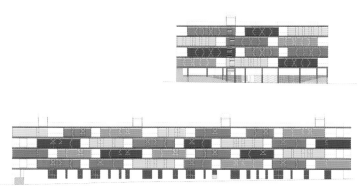

立面图 ELEVATIONS

© David Frutos

© David Frutos

卡拉班切尔住宅楼，Amann-Cánovas-Maruri，西班牙，马德里，2009年

这栋建筑的设计理念并非来源于传统住宅楼，而是采用了宽度极小的穿孔板材。建筑外部是由金属构成的，因此可以作为通风立面。建筑整体色彩斑斓，吸引眼球，就像是一系列车身的集合，住户可自主选择"车身"的金属颜色。

Carabanchel Housing, Amann-Cánovas-Maruri, Madrid, Spain, 2009

The proposal is not built from the review of the traditional housing block but from the attributes of the slab of minimum width perforated with through holes. The exterior body is constructed of metal, therefore acts as a ventilated façade. The building is striking with many bright colors, like an ordered set of car bodies whose metallic colors are the choice for users.

材料：金属

Material: metal

虚幻的蛊惑
PSYCHEDELIC DELUSION

 猩红 Crimson

 纯黑 Black

雪白色 Snow

草间弥生用一种对待种子的心意照顾着这些色彩绚丽的波点，让它们无限的繁殖，生长出一个虚幻的世界。而她就在她所创造的世界中等待着你，用斑点诱惑你与她一起疯狂。草间弥生这位被视为日本现存最伟大的"怪婆婆"艺术家，专注地用它标志性的波点语言与世界保持着对话。

Yayoi Kusama takes good care of the round dots as they were seeds, letting them reproduce indefinitely to create an illusory world. And she is waiting for you in this world and tempting you to go crazy with her. Yayoi Kusama, who is nicknamed "odd granny" and regarded as the greatest existing artist in Japan, has been attentively keep a dialogue with the world with his signature dot language.

① 通往新空间的路标，草间弥生，美国，纽约，2004年
Guidepost to the New Space, Yayoi Kusama, New York, USA, 2004

② 红南瓜，草间弥生，日本，直岛，2006年
Red Pumpkin, Yayoi Kusama, Naoshima, Japan, 2006

③ 在树上攀升的波尔卡圆点，草间弥生，新加坡双年展，2006年
Ascension of Polkadots on the Trees, Yayoi Kusama, Singapore Biennale, 2006

© Courtesy of Yayoi Kusama Studio Inc. ①

© Courtesy of Yayoi Kusama Studio Inc. ②

© Courtesy of Yayoi Kusama Studio Inc.

猩红 Crimson

 纯黄 Yellow

 亮粉红 Light Pink

 纯蓝 Blue

 闪光深绿 Lime Green

 雪白色 Snow

 纯黑 Black

① 圆点沉迷，草间弥生，英国，伦敦，2009年
Dots Obsession, Yayoi Kusama, London, Britain, 2009

② 圆点沉迷—新世纪，草间弥生，法国，第戎，2000年
Dots Obsession-New Century, Yayoi Kusama, Dijon, France, 2000

③ 圆点沉迷—白天，草间弥生，美国，华盛顿，2008年
Dots Obsession-Day, Yayoi Kusama, Washington, USA, 2008

④ 圆点沉迷—爱的化身，草间弥生，德国，慕尼黑，2007年
Dots Obsession-Dots Transformed into Love, Yayoi Kusama, Munich, Germany, 2007

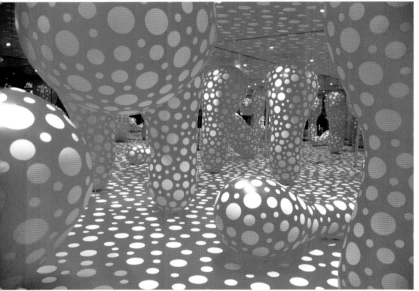

© Courtesy of Yayoi Kusama Studio Inc.　①

© Courtesy of Yayoi Kusama Studio Inc.　②

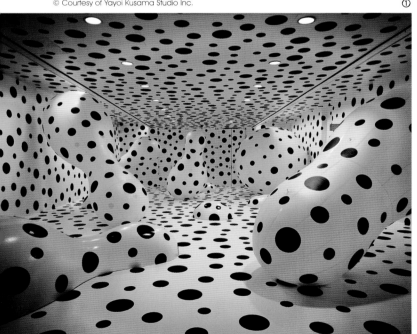

© Courtesy of Yayoi Kusama Studio Inc.　③

© Courtesy of Yayoi Kusama Studio Inc.　④

猩红 Crimson

 暗淡灰 Dim Gray

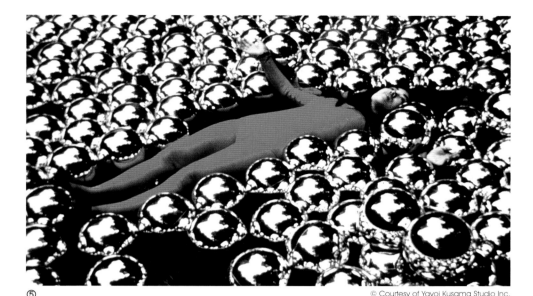

© Courtesy of Yayoi Kusama Studio Inc.

© Courtesy of Yayoi Kusama Studio Inc.

当望着这些波点时，会有一种强烈的错觉和在浩瀚斑点中分不清南北的疑惑。各种颜色排列组合着，纯粹而耀眼，没有丝毫让人倦怠的艳俗，而是有种身心享受的放松。有人说"草间弥生不知在哪面墙上钻了个洞，窥知了造物主的某个手势或背影，她从此就寄居在这面墙上，在两个世界间来回顾盼"。是这样的，她展示给人们她眼中的波点，带来了一个与她一起幻想和迷狂的虚幻世界。

When you gaze at these dots, a strong illusion and confusion will rise to get you lost in the unlimited dots. Different colors were arranged to offer a pure and dazzling image. No a bit frippery will exhaust you. Instead, it allows you to relax mentally and physically. "Yayoi Kusama dug a hole on some wall to take a furtive glance at a gesture and figure of God. She has been living there since then and casts glances about between the two worlds, " some people said. In such a way, she shows her dots with people and brought a visionary world for people to fantasize about with her.

⑤ 那克索斯的花园，草间弥生，意大利，威尼斯双年展，1966年
Narcissus Garden, Yayoi Kusama, Venice Biennale, Italy, 1966

⑥ 无限镜屋，草间弥生，美国，纽约，1965年
Infinity Mirror Room, Yayoi Kusama. New York, USA, 1965

■绿，帛青黄色也。——《说文》■大自然通过颜色来帮助我们识别生命形态、健康状况、果实成熟、四季变换。■自然生机的代表色即为绿色，象征新的生长、青春、活力、新老交替、平衡、和谐、康复。■在自然中汲取色彩的方法需要注重技巧和细节，其结果总是立竿见影，这种颜色方法鼓励人们构思出亲切的、生机勃勃的环境，可以与室外的树木、水土和天空构成的自然环境相媲美。■绿色对眼睛来说是最放松的颜色，也是自然中唯一融合了暖黄色和冷蓝色的中性颜色。由于对神经系统有镇静作用，绿色适用于休闲放松的场所。各种派生色适用于墙壁、顶棚和地板。

■ Green is made by mixing blue and yellow together. ■ By means of colors Nature helps us to identify life forms, physical conditions, maturity of fruits and seasonal changes. ■ Green is the representative color of the nature and vitality, and it stands for growth, youth, energy, supersession of the old by the new, balance, harmony and recovery. ■ Absorbing colors from nature, with special attention on skills and details, will always produce an immediate effect. Getting inspiration from nature encourage people to create an amiable, vigorous environment which can compare favorably with the natural environment composed of trees, water, soil and the sky. ■ Green is the most relaxing color for eyes and the only intermediate color mixing warm yellow and cool blue. It is suitable for leisure places since it can calm down active nerves. All its derivative colors are suitable for walls, ceilings and floors.

绿之生机
GREEN
099

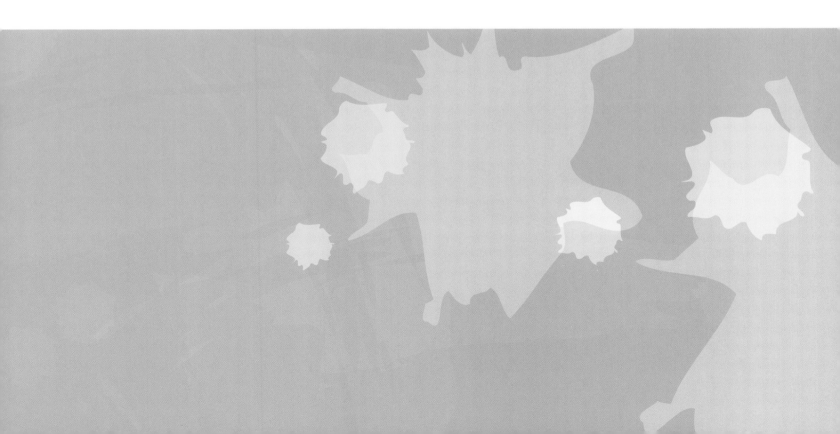

生机盎然的森林
VITALITY OF FOREST

 硬木色 Burly Wood

 绿黄色 Green Yellow

 闪光绿 Lime

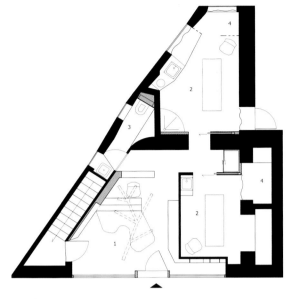

1 招待区　　1 Reception
2 身体护理室　2 Body Treatments Room
3 卫生间　　3 Toilets
4 储藏室　　4 Storeroom　　　平面图 PLAN

© Trust in Design

© Trust in Design

B.institut美容院，信托设计，法国，瓦讷，2008年

B.institut是位于法国瓦纳市的一所美容院。设计师力求创造一个带有禅宗意境的温馨场所。为此，设计师需要为美容院树立新的品牌形象。在内饰设计上，设计师采用了山毛榉为家具材料，地面则以绿色树脂铺设而成。墙壁上贴有电脑膜切装饰图案。设计团队对房间内的家具和照明进行了巧妙的设计，在树立品牌形象的同时，也创造了一个具有亲和力的空间。

B.institut Beauty Parlor, Trust in Design, Vannes, France, 2008

B.institut is an organic beauty parlor in Vannes. The designers would like to create a welcoming place with a Zen and peaceful spirit. The designers had to create a new brand identity for this beauty parlor. The designers furnished the interior with beech units and coated the floor in green resin. Computer-cut stickers decorated the walls. The design team created a welcoming place while creating the identity of the brand, with an organic design for the furniture and the lighting.

材料：内部 装配榉木家具
　　　地板 覆有绿色树脂
　　　墙壁 装饰有电脑切割贴纸

Materials: Interior: furnished with beech units
　　　　　Floor: coated in green resin
　　　　　Walls: decorated with computer-cut stickers

立面图 FAÇADE

双色太阳镜
DUAL COLOR SUNGLASSES

 黄绿色 Yellow Green

 森林绿 Forest Green

 紫罗兰 Violet

 雪白色 Snow

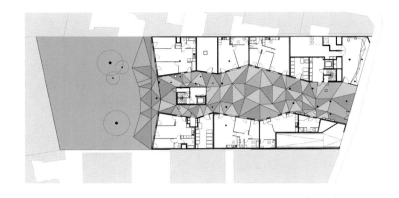

总平面图 SITE PLAN

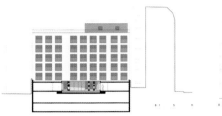 

立面图 ELEVATIONS

© Cyrille Thomas

社会住房:63套公寓,Emmanuel Combarel Dominique Marrec architectes (ECDM),法国,巴黎,2008年

这座住宅楼位于街道的边缘,并留有大块的开阔空间,延绵至一座标志性的花园。建筑的正立面由色调深浅不一的绿色玻璃构成。庭院内铺装着矿物质地板,布满了绿色的浮雕图案,犹如万花筒一般。项目的四个立面呈现了两种色彩,并根据具体的情况略有变化。所有立面均设有宽阔的门窗,并连接着大型的阳台或露台(根据各自的朝向)。镶嵌在立面上的彩色玻璃像太阳镜一样,保护着建筑表皮。整座建筑流淌着汩汩生机,生命的色彩盈盈若溢。

Social Housing: 63 Apartments, Emmanuel Combarel Dominique Marrec architectes (ECDM), Paris, France, 2008

The accommodation is concentrated on the street side which leaves a wide open space that reaches the landmark garden. The front façade is composed of colored glass in different shades of green. The mineral courtyard is treated like a kaleidoscope colored in a cameo of green. The project presents 2 colors and 4 specific façades conceived to respond to very specific conditions, all characterized by wide windows, opening onto large terraces or balconies (depending on their orientation) and protected by colored glass which is treated like sunglasses. The entire building is filled with vigor and vitality as if the color of life were about to overflow.

© Cyrille Thomas

材料:玻璃

Material: glass

© Cyrille Thomas

动物的天堂
ANIMALS' HEAVEN

 暗绿色 Dark Green

 橄榄褐色 Olive Drab

 暗海洋绿 Dark Sea Green

 中海洋绿 Medium Sea Green

 草绿色 Lawn Green

 绿黄色 Green Yellow

 纯黄 Yellow

 橄榄褐色 Olive Drab

动物庇护所，Arons en Gelauff architecten，荷兰，阿姆斯特丹，2007年

驱车至荷兰阿姆斯特丹市郊，会看到一栋彩色丝带般的建筑，外观由深浅不一的绿色随机排列而成，绵绵的绿色如屏障一般与城市边缘的自然之绿浑然一体。18种绿色交相呼应，错落有致，精致而不张扬，典雅而不失活泼情趣。两个三角形的核心平面，由公共管理区及办公空间连接起来。流线造型的入口及大厅成为了设计的点睛之笔。入口位置的设计打破了直线型延续外墙的呆板，整个建筑外围变得生动起来；同时，道路在暖暖的色彩包围中逐步深入，为与动物们的接触创建了一种情感上的铺垫。

Animal Refuge Center, Arons en Gelauff architecten, Amsterdam, Netherlands, 2007

When driving in the suburb of Amsterdam, a colorful ribbon like building will jump into our sight. Its façade consists of randomly arranged colors of different hues. The color of the green "screen" smoothly merged into the natural color in the urban fringe. 18 different hues of green color are properly distributed and corresponded with each other, thus creating a graceful yet vivid appearance. Two triangle core planes are connected by the public and office spaces, with fluid entrance and foyer being the highlight of the design. The entrance is designed to break down the stiffness of the continuous wall, vivifying the outside of the building. Surrounded by the warm colors, the paths stretch into the courtyard, thus making emotional preparation for the communication with animals.

材料：金属面板、玻璃

Materials: metal panels, glass

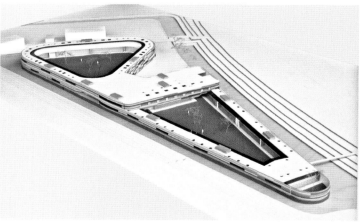

模型 MODEL

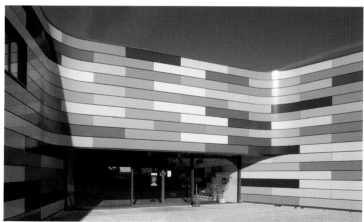

© Luuk Kramer

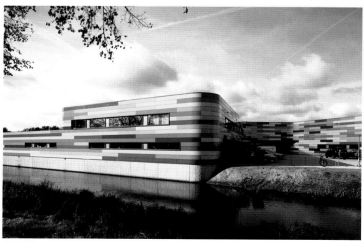

© Luuk Kramer

© Luuk Kramer

© Luuk Kramer

神奇小人国
THE MYSTERIOUS LILLIPUTIAN

 猩红 Crimson

 闪光深绿 Lime Green

 亮粉红 Light Pink

 淡绿色 Light Green

 亮菊黄 Light Goldenrod Yellow

 银灰色 Silver

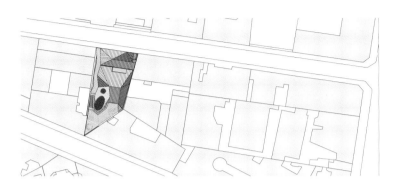

总平面图 SITE PLAN

© Courtesy of BP Architectures

© Courtesy of BP Architectures

法国皮卡尔幼儿园，BP Architectures，法国，巴黎，2010年

皮卡尔街道位于蒙马特高地脚下，是一个具有浓厚巴黎风情的道路，直直的街道，密密的建筑。幼儿园位于有序、单一的灰白色石墙之间，就像《格列佛游记》中的小人国世界一样别具特色。仿佛一粒稻米落入砂子之中，这个幼儿园结束了这条街区传统的城市风格，成为本区域的一个标志性的建筑。在构思之初，它就是一个锌板折成的体量，将其展开，折痕和面形成连续不断的墙面和屋顶。宏观上看，仿佛是屋顶置于地面，而上方全部成为阁楼，形成一处神秘、奇幻、有趣的乐园，所有小孩甚至大人都会喜欢。原色色块的运用，是该设计的点睛之处，色块将门窗有趣地强调出来，又创造了一种独有的关联。高矮胖瘦的彩色"门窗"排排站，带来奇幻的童话气息。

Picard Nursery in Paris, BP Architectures, Paris, France, 2010

Located at the foot of the Butte Montmartre, rue Picard seems like a very Parisian street, with its dead straight alignment and high density. And yet in the middle of this ordered, monotonous, monochrome urbanity with its off-white stone-colored walls, a little nursery emerges like a Lilliputian in Gulliver's world. Like a grain of sand, it manages to seize up the urban machine and provide the area with a real reference point. Before being an architectural design, the building is a volume: a continuous expanse of zinc combining the vertical walls of the façades and the folds of the roof. The building is, in effect, a roof placed on the ground, as if all that was left of the house was its attic, a mysterious, scary and exciting world that all children (and adults) love. The use of primary colors is the highlight of the design. With the involvement of the colors, the doors and windows are emphasized in a pretty interesting way and a unique connection was therefore created. The "doors and windows" in different sizes and colors stand in rows, which created a fantastic ambience of fairy tales.

材料：彩色金属面板、玻璃

Materials: colored metal panels, glass

© Courtesy of BP Architectures

绿意盎然
FULL OF GREEN

 蜜色 Honeydew

 黄绿色 Chartreuse

 中宝石碧绿 Medium Aquamarine

荷兰代尔夫特理工大学学生公寓，Mecanoo Architecten，荷兰，代尔夫特，2009年

这3座四面泛光的公寓楼均有6个楼层，共设有186个居住单元，并与一个牧场毗连。3个立面均为黑色的砌体结构，并装有紧压式连接构件和铝框滑动门。居住者可以透过大型窗户俯瞰周边景色的全貌。在靠近牧场的一侧，这3座公寓楼呈现了富有层次的绿色立面。立面由复合材料面板构筑，并装有管架以供攀缘植物在春天生长攀爬。这些攀援植物会在秋季落叶枯萎，复合材料面板上翠绿的色彩则会随之显现出来。这3座建筑仿佛覆盖着草毡和石板一般，颇具自然风格。

Student Housing, the Technical University of Delft, Mecanoo Architecten, Delft, the Netherlands, 2009

The three omni-sided residential towers have a total of 186 residential units, each six storeys high bordering a pasture. Three façades are finished in dark masonry with squeezed joints and aluminium framed sliding doors. Large windows offer panoramic views of the surroundings. On the pasture side, the three buildings display a layered green façade of composite plates in combination with a tubular frame which allows climbing plants to grow in the spring. In the fall, when the climbers lose their leaves, the plant motif and green color of the composite plates become visible. The three buildings are enveloped in a grass carpet strewn with natural and flagstones.

材料：复合材料面板、植物

Materials: composite panels, plants

© Christian Richters

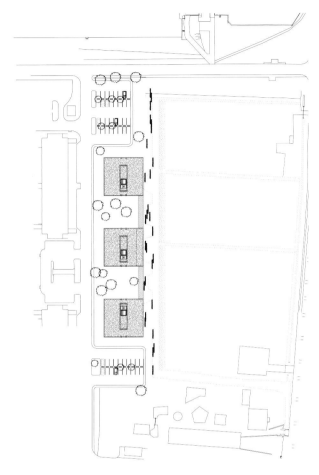
总平面图 SITE PLAN

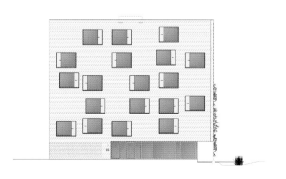
立面图 FAÇADES

律动的生机
DYNAMIC VITALITY

 银灰色 Silver

 柠檬黄 Lemon Yellow

 绿黄色 Green Yellow

 闪光深绿 Lime Green

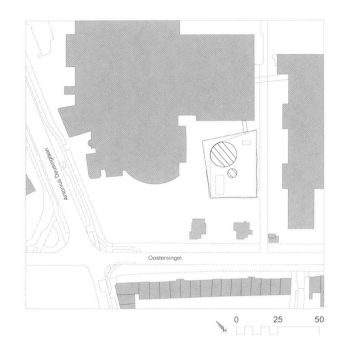

总平面图 SITE PLAN

© UNStudio

© UNStudio

研究实验室，UNStudio，荷兰，格罗宁根，2008年

建筑外表由平滑的纵向铝板构成，铝板在一些位置向外弯折成弓形，从而形成了纵向的波浪，由于视角不同，形成了或开放或封闭的视野。另外，铝板向侧面弯曲，并涂有明亮的色彩，其视觉效果使得严肃的外观更加富有生气。建筑较低楼层的铝板为黄色，向上到顶层逐渐变化成为绿色，且通过渐变、高差和图案呈现出迷人的欧普视幻效果。

Research laboratory, UNStudio, Groningen, the Netherlands, 2008

The façade is composed of flat, vertical aluminum slats, which, in some places, are twisted outwards in bowed forms. Thus, tall vertical undulations are generated, which present an open or a closed aspect depending on the angle under which they are viewed. This optical effect enlivens the restrained façades, which is made even stronger by the application of bright colors to the flat surfaces that are created by twisting the aluminum slats sideways. On the lower level the color yellow is used, which gradually changes to green towards the top of the building. Through color gradient, height difference and images, an amazing optical effect was produced.

材料：混凝土、铝板

Materials: concrete, aluminum slats

© Christian Richters

七彩童年
COLORFUL CHILDHOOD

 硬木色 Burly Wood

 纯黄 Yellow

 橄榄 Olive

 闪兰色 Dodger Blue

 深粉红 Deep Pink

立面图 FAÇADES

© Courtesy of BP Architectures

© Courtesy of BP Architectures

法国埃皮奈幼儿园，BP Architectures，法国，埃皮奈苏塞纳尔，2010年

幼儿园共分为5个部分，都以"童年"为主题，但是它们各具独自的色系、构造形式和入口。建筑被绿色的植被分割成一个个小单元，与主干道形成直角。室内根据活动功能不同，各部分镶板屋顶坡度也随之变化。而脊梁的高度也由室内活动的需求而决定。幼儿园这一功能决定了建筑色彩斑斓的外立面，与周围白色调的建筑群形成了鲜明对比，也为该地区涂抹上一笔明亮的色彩。幼儿园附近栽种着一组高大的松树，顺着斜坡缓缓地延伸到Yerres河边，面朝河岸能眺望远处的乡村风光。

Epinay Nursery School, BP Architectures, Epinay-sous-Senart, France, 2010

The project consists of 5 entities, all themed by childhood but each one with a distinct color system, configuration and access. The project developed into a group of small units at right angles to the main access road and alternating with strips of vegetation. Each unit has a paneled roof whose slope differs according to the activities underneath. The height to ridge beam and therefore the resulting available internal space are linked to the room's importance. The function of the kindergarten results in a colorful façade, which contrasts sharply with the surrounding buildings and brings a shade of bright color to this area. The land features a clump of tall pine trees, an extensive grassy area that slopes gently down to the Yerres River and a view of the distant countryside beyond the opposite bank of the river.

材料：木材、玻璃

Materials: wood, glass

© Courtesy of BP Architectures

© Courtesy of BP Architectures

© Courtesy of BP Architectures

© Courtesy of BP Architectures

© Courtesy of BP Architectures

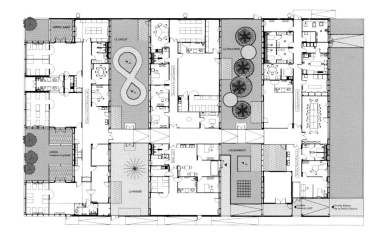
平面图 PLAN

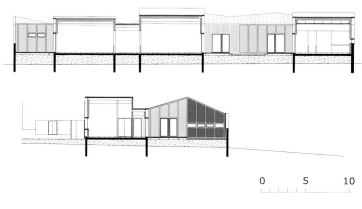
剖面图 SECTIONS

法国埃皮奈某幼儿园，BP Architectures，法国，埃皮奈苏塞纳尔，2010年

这5个部分都有自己专属的室外花园。其中托儿所的3个部分设计相似：面向室外游戏区的朝南活动室，有彩色的天井采光。充满生机的色条同样运用在室内的墙面与家具上，让孩子们无时无刻都能感受到色彩的魅力。

Epinay Nursery School, BP Architectures, Epinay-sous-Senart, France, 2010

Outside, each project entity has its counterpart in its adjoining garden. The nursery's three sections have similar layouts, a south-facing activity room opening onto an outside play area and lit by colored light wells. His corridor allows movement from areas open to the public to the most private parts reserved just for the children. The same fascinating colors also appear on wall surfaces and furniture so that children can enjoy the charms of colors at any time.

材料：木材、玻璃

Materials: wood, glass

硬木色 Burly Wood

 纯黄 Yellow

 橄榄 Olive

© Courtesy of BP Architectures

悄然绽放的花朵
A SILENTLY BLOOMING FLOWER

 弱绿色 Pale Green

 浅海洋绿 Light Sea Green

 硬木色 Burly Wood

© Fabio Mantovani

© Fabio Mantovani

© Fabio Mantovani

巴巴爸爸幼儿园,CCD studio,意大利,摩德纳,维尼奥拉,2009年

该项目旨在于各个方面都想体现出可持续性,并期待建筑本身也将这种价值观传递给了孩子们。和谐中性的色彩与建筑的周围绿色的植被融合在了一起,减少了建筑所带来的视觉冲击。木制的屋顶之上覆盖着使室内保持舒适绿色的甲板,具有良好的隔热效果。宽大的玻璃窗系统在一天中的任何时候都可以对太阳光线进行隔滤,以保持室内温度。绿色的外立面契合了可持续这一主题,同时使建筑看上去清新宜人。当夜色降临时,灯光透过彩色玻璃窗射在室外的绿色草坪上,似花朵悄然盛开,芬芳灿烂。

Kindergarten Barbapapà, CCD studio, Vignola, Modena, Italy, 2009

The project aimed to express the sustainable theme in all aspects and to transmit this value to the children. In order to protect the children and reduce the visual impact of the volume, the kindergarten was surrounded by green vegetation. The green deck ensures to maintain a good thermal insulation, to preserve the environmental comfort with a terrain package placed on top of the roof's wood structure. The large glass system filters sunlight at any time of the day to maintain the heat inside. The sustainable theme integrates into the green façade, which makes the building clear and pleasant. When night falls, the light cast onto the green grassland through the stained glass windows, and the building becomes a blossoming flower.

材料:木材、玻璃

Materials: wood, glass

© Fabio Mantovani

© Fabio Mantovani

纯黄 Yellow

 闪兰色 Dodger Blue

 重褐色 Saddle Brown

巴巴爸爸幼儿园，CCD studio，意大利，摩德纳，维尼奥拉，2009年

　　室内空间中的家具和色彩具有很强的趣味性，满足了孩子的好奇心，也增强了整个空间的饱和感。颜色、材料和特别的设计形状体现出了整体的设计思路，同时也体现出了整个建筑的现代感。

Kindergarten Barbapapà, CCD studio, Vignola, Modena, Italy, 2009

The furniture and colors in the interior spaces are interesting and therefore satisfy children´s curiosity. The color saturation of the entire space is also promoted. Colors, materials and the special shape used in this building express the general design concept and reveal the modernity of the building.

材料：木材、玻璃、涂料

Materials: wood, glass, paint

© Fabio Mantovani

深橙色 Dark Orange

 金色 Gold

 硬木色 Burly Wood

© Fabio Mantovani

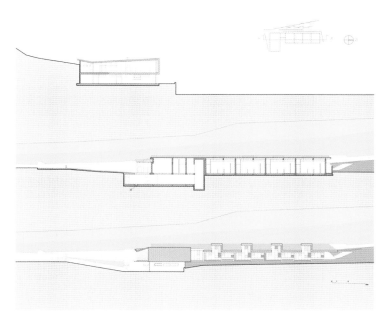
剖面图 SECTIONS

© Fabio Mantovani

1 通道
2 入口
3 等候区
4 办公管理区
5 洗手间
6 亲子中心
7 楼梯和电梯
8 公共区域
9 工作室
10 教室
11 天然活动场所
12 紧急逃生通道

1 Trail Access
2 Entrance
3 Waiting
4 Office Administration
5 Rest Room
6 Center For Childrens And Parents
7 Staircase And Elevator
8 Common Area
9 Work Room
10 Classroom
11 Natural Arena
12 Emergency Escape

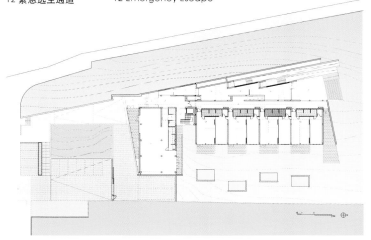
一层平面图 GROUND FLOOR PLAN

© Fabio Mantovani

绿之生机 | GREEN | 119

杨树林的守卫
POPLAR GROVE'S ESCORT

 绿黄色 Green Yellow

 草绿色 Lawn Green

 闪光深绿 Lime Green

© José Manuel Cutillas

© José Manuel Cutillas

2C住宅，Vaíllo & Irigaray + Eguinoa，西班牙，纳瓦拉，2010年

这座建筑的绿色"栅栏"立面模仿了河床和灌溉水渠两旁栽种着成排的杨树，这些蜿蜒列立的杨树的使远处的河流看起来斑驳朦胧，这是大自然干涸的景象中显现的生命征象。Cañete河穿过建筑场地，并流向南方。这座住宅充分利用了这种地理优势，在北侧封闭，向南方开敞。曲折延绵的立面和可以上下移动的条栅成就了建筑多变的外观，呈现了如行行杨树般的蜿蜒景象。里外不同层次的绿色似一座小小的森林，充满整个视线，且步移景异，乐趣无穷。

2C Houses, Vaíllo & Irigaray + Eguinoa, Navarra, Spain, 2010

As many others, all over Ribera's ("River Bank") landscape, meandering poplar rows normally escort the entire length of river beds and irrigation ditches. Far in the distance, river beds are spotted thanks to those winding rows: it is a gesture offered by nature as a sign of life in an arid, dry scene. The lot, traversed by Cañete River, slopes down to the south… advantage of this circumstance is taken to arrange a house that is sheltered from the north and exposed to the south…The houses suggest a changing image, prompted by the façade crooked geometry and the latticework vertical movement, that offer a meandering scenery, as poplar rows do…Different layers of green colors look like a small forest filling the eyes. The scenery constantly changes with people's movement, bringing unlimited fun.

材料：玻璃

Material: glass

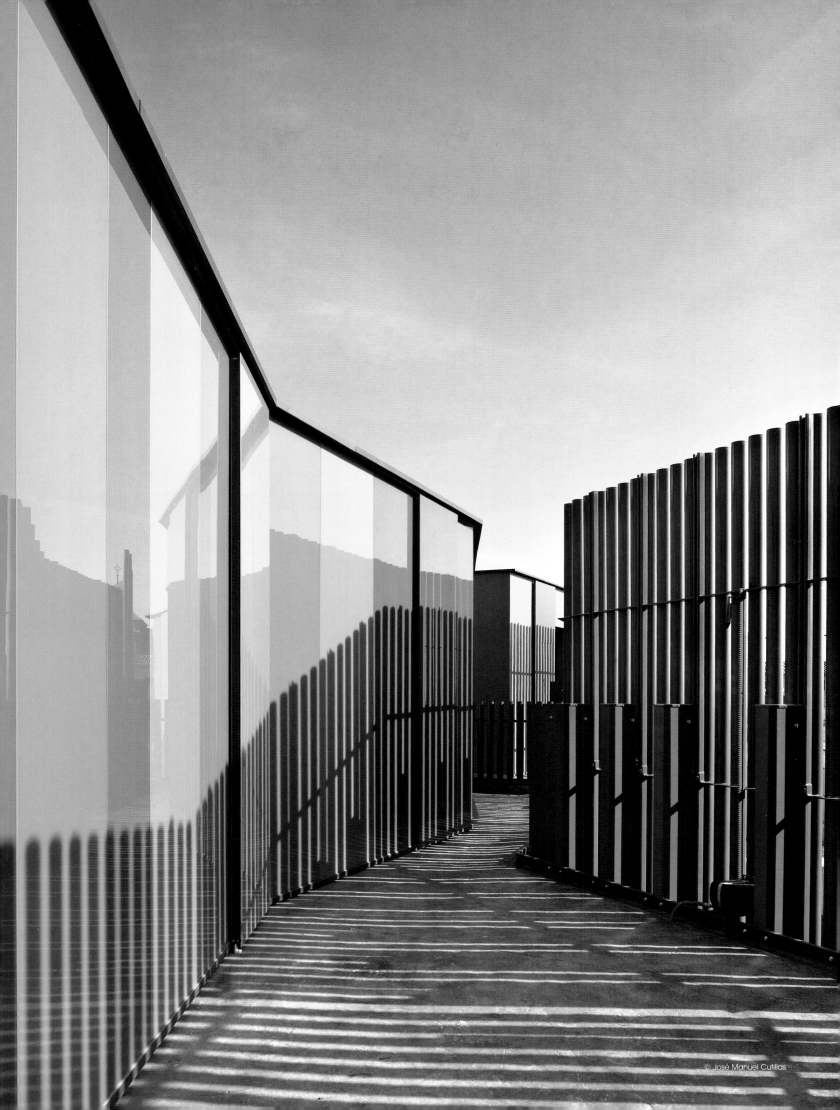
© José Manuel Cutillas

绿铅笔
THE GREEN PENCILS

 亮菊黄 Light Goldenrod Yellow

 淡绿色 Light Green

 闪光深绿 Lime Green

 茶色 Tan

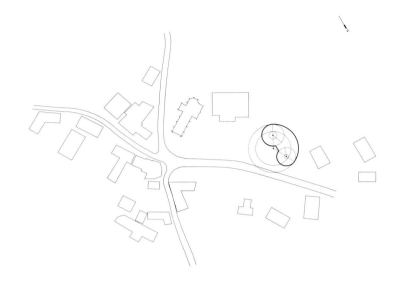

总平面图 SITE PLAN

© Jérome Ricolleau

© Jérome Ricolleau

幼儿中心，Tectoniques，法国，安省，圣迪迪埃德福尔芒，2011年

幼儿中心的设计灵感来源于童年的世界，使用圆曲线向内弯曲，同时延伸融入景观，使建筑与自然完美融为一体。圆形的形状让人们联想到糖果、蛋糕或游戏，而曲线则象征着儿童时期所需要的柔软性，同时能够使视线沿建筑物的轮廓平滑地移动。建筑覆有木制外层，由巨型"铅笔"形的道格拉斯冷杉板条固定到三层的黑色落叶松板上。"铅笔"边缘都被漆成深浅不一的绿色，被命名为"绿铅笔"，成为当地的象征。建筑外墙营造出一种动感氛围，吸引人们沿着建筑物漫步，就像是摄影机移动拍摄。建筑的表皮不仅具有美学效果，它还能起到防晒作用，各"铅笔"的厚度根据角度调整。

Infants' Center, Tectoniques, Saint-Didier-de-Formans, Ain, France, 2011

Drawing its inspiration from the world of childhood and using round curves to bend inwards upon itself and at the same time stretch out to embrace the landscape, the Center merges harmoniously into its natural environment. Its round shapes that are reminiscent of sweets, cakes or games, and the curves suggest the softness required during childhood and allows the eye to slide smoothly along the contours of the building. The building is covered with a skin made out of giant pencil-type Douglas fir slats fixed onto black coated three-fold larch boards. The "pencil" edges are painted in shades of green and are the symbol of the site that was recently named "The Green Pencils". The outer jacket creates a kinetic effect and makes one want to glide along the building as with a cinema travelling shot. The outer skin is not only aesthetic: it also serves as a sunscreen and the thickness of each pencil piece is adjusted according to the angle.

材料：木材

Material: wood

© Jérome Ricolleau

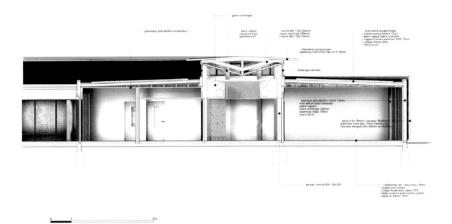

剖面图 SECTION

© Jérome Ricolleau

© Jérome Ricolleau

金色 Gold

 黄土赭色 Sienna

 雪白色 Snow

 沙棕色 Sandy Brown

 纯黑 Black

1 主广场	1 Parvis d'accueil
2 门廊	2 Sas d'entrée
3 门厅	3 Hall
4 活动室	4 Salles d'activités
5 寝室	5 Salons de sommeil
6 餐厅	6 Restaurant
7 办公室	7 Bureaux

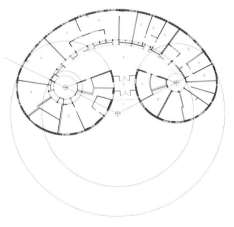

平面图 PLAN

© Jérome Ricolleau

幼儿中心,Tectoniques,法国,安省,圣迪迪埃德福尔芒,2011年

室内延续了室外的色彩,明快的黄色、绿色洋溢着生命的活力。内部空间也反映出外部形状,整个项目围绕着一个巨大的新月形走廊展开布局,中庭及室内大部分采用自然采光。走廊的设计使建筑布局紧凑,同时保持了与外部的联系。走廊还将项目的几个圆圈结构连通,象征着外部与内部之间的中间空隙。这是一个公共空间,供儿童、家长以及工作人员使用。

Infants' center, Tectoniques, Saint-Didier-de-Formans, Ain, France, 2011

The outer shape gives the imprint of the inner space. The project is organized around a vast croissant-shaped hallway, lit for the most part by natural light and that some call the atrium. The hallway confers compactness on the building while maintaining a link with the outside. It also links the geometric circles of the design and symbolizes the intermediate space between the outside and the inside: it is the space that is shared by all; children, parents and staff alike.

材料:木材

Material: wood

© Jérome Ricolleau

绿光森林
GREEN FOREST

 亮黄色 Light Yellow

 春绿色 Spring Green

 闪光深绿 Lime Green

 暗淡灰 Dim Gray

 象牙色 Ivory

南立面图 SOUTH FAÇADE

北立面图 NORTH FAÇADE

东立面图 EAST FAÇADE

西立面图 WEST FAÇADE

© Crepain Binst Architecture

© Crepain Binst Architecture

Infrax 办公楼，Crepain Binst Architecture，比利时，托尔豪特，2010年

典型的生态和可持续性理念在建筑中表现得淋漓尽致，若干条纤长的混凝土"树干"擎起了宽阔的绿色"树冠"，隐喻森林的形象。设计的核心亮点是创新型的墙壁，它无论是在视觉上还是在象征意义上都体现了"绿色"的特征。建筑上部封闭式的体块与开敞通透的底层形成了鲜明的对比。建筑的外表皮由呈现三种颜色和三种透明度的丝网印刷玻璃板构成，并利用这些玻璃板拼接成了马赛克图案的墙壁。其别具一格的构造无疑强化了建筑的表现力和动态感。

Offices Infrax West, Crepain Binst Architecture, Torhout, Belgium, 2010

The characteristic ecological and sustainable values were given expression in this project by using a metaphor for a forest in which a host of slender concrete trunks were topped with a broad green "crown". The central pillar of the design is the creation of an innovative wall that is both literally and metaphorically green. The closed character of the upper building is in contrast to the transparency and openness on the ground floors. The outer skin consists of screen printed glass panels in three colors and three degrees of transparency. The mosaic formed by all these elements and the wall's ingenious construction have made an unmistakable contribution to the building's expressive and dynamic character.

材料：混凝土、丝网印刷玻璃

Materials: concrete, screen printed glass

© Crepain Binst Architecture

弱绿宝石 Pale Turquoise

 灰色 Gray

 浅灰色 Light Gray

© Crepain Binst Architecture

© Crepain Binst Architecture

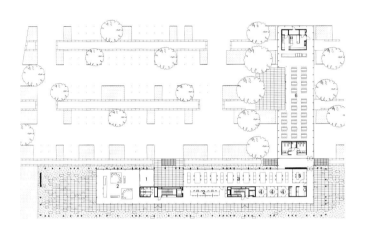

1 入口	1 Entrance
2 前台/接待处	2 Reception / Front Office
3 景观设计办公室	3 Landscape Office
4 会议室	4 Meeting Room
5 办公室	5 Office
6 午餐区	6 Lunch
7 储藏室	7 Storage
8 工业场所	8 Technical Space
9 服务/制图场所	9 Server / Plotter Space

平面图 PLAN

Infrax办公楼，Crepain Binst Architecture，比利时，托尔豪特，2010年

这种技术性创新与毫无阻断的办公空间结合在了一起：3.4 m的净空高度和跨度16 m的无柱设计使空间拥有了极佳的通风和灵活性。建筑内部采用低调的的装饰风格，以白色为主色调，地板和隔声屏则采用了温暖的灰色。精心挑选的黑色则用于凸显建筑的辅助功能和垂直流通空间。

Offices Infrax West, Crepain Binst Architecture, Torhout, Belgium, 2010

This technical ingenuity has been combined with unbroken space in the offices: A large free height of 3.4 m and a column-free span of 16 m have made the space airy and extremely flexible. The décor was deliberately kept restrained, using white as the main color and warm shades of gray for the flooring and acoustic screens. A few carefully chosen black accents emphasize secondary functions and vertical circulation in the building.

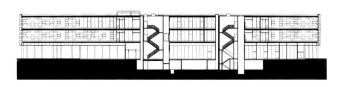

剖面图 SECTIONS

材料：混凝土、玻璃

Materialss: concrete, glass

© Crepain Binst Architecture

© Crepain Binst Architecture

绿之生机 | GREEN | 129

水的世界
THE WORLD OF WATER

 薄荷奶油 Mint Cream

 烟灰色 Smoke Gray

浅海洋绿 Light Sea Green

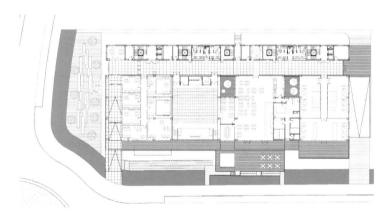

总平面图 SITE PLAN

© David Frutos

© David Frutos

Pita and Tecnova总部大楼，Ferrer Arquitectos，西班牙，阿尔梅里亚，2012年

该项目是首座建于公园内的建筑，建筑的可持续性设计反应出了公园和即将入驻的公司的愿景。朝南的外立面安装横向的百叶窗，使得内部空间免受光线的直射。百叶窗后是玻璃立面，保证了办公室和实验室享有充足的自然光线。建筑淡绿色的外立面与公园的绿色主题交相呼应，百叶窗构成的层层阴影，像树叶斑驳，清新自然。

Pita and Tecnova Headquarters, Ferrer Arquitectos, Almería, Spain, 2012

The buildings were the first to be completed in the park and were sustainably designed to reflect the ideals of the park and the businesses that would work there. The south-facing exterior is covered in a large array of horizontal louvers to shield the interior from the direct sun. Stepped in from the louvers is a fully glazed façade that lets in lots of natural daylight into the office and lab spaces. The light green façade of the building responds to the green theme of the park and the shadows made by the louvers are just like mottled leaves, fresh and natural.

材料：玻璃

Material: glass

© David Frutos

© David Frutos

薄荷奶油 Mint Cream

 烟灰色 Smoke Gray

 浅海洋绿 Light Sea Green

 绿宝石 Turquoise

© David Frutos

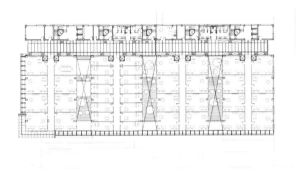
平面图 PLAN

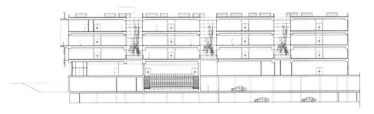
纵向剖面图 LONGITUDINAL SECTION

© David Frutos

Pita and Tecnova总部大楼，Ferrer Arquitectos，西班牙，阿尔梅里亚，2012年

室内的空间以如水般的淡绿色调为主，层层浅绿色的透明磨砂玻璃地板使建筑显得更为通透光亮。在这水晶般的空间中，生长着墨绿色攀爬植物的跃层庭院散布开来。整个建筑恍若一个水的世界。每一个楼层均设有一个悬伸于周边景观之上的会议室，海湾和Gata Cape自然公园的景色一览无余。

Pita and Tecnova Headquarters, Ferrer Arquitectos, Almería, Spain, 2012

The inner space is tinted in shade of light green, like water diffuses. Layers of green transparent ground glass lighten the entire building. The scattered courtyards with dark-green climbing vegetation provide a crystal-like space. The whole building looks like a water world. Each floor contains a meeting room that protrudes out over the landscape, offering views of the bay and the Gata Cape Natural Park.

材料：玻璃

Material: glass

绿草艾艾
GREEN GRASS

 闪光深绿 Lime Green

 军兰色 Cadet Blue

 暗瓦灰色 Dark Slate Gray

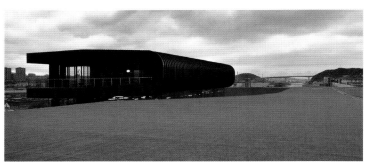
© Aitor Ortiz

© Aitor Ortiz

© Aitor Ortiz

总平面图 SITE PLAN

IDOM总部大楼,Javier Pérez Uribarri, ACXT Arquitectos,西班牙,毕尔巴鄂,2011年

从外部看,建筑的屋顶像是被一条绿色的地毯所覆盖,所有的空调设备都被隐藏起来,从而减少了不必要的声音和视觉影响。设计者采用了水平的设计方案,在建筑美感、造价和维护方面取得了最佳平衡。在建筑表皮上,不透明的部分采用复合铝材和岩棉,而柱间的空间则采用了高性能玻璃和铝制框架构成的幕墙。板条和立面之间还设有一个平台,用于从外部清洁玻璃。绿色的条纹立面与草坪融为一体,仿佛是一座被绿色覆盖的小山丘耸立在岸边,生机勃勃。

IDOM Headquarters, Javier Pérez Uribarri, ACXT Arquitectos, Bilbao, Spain, 2011

Outwardly, an imaginary green carpet has been designed as if simply placed over the roof, hiding all air conditioning units. As a result, the sound and visual impact are reduced. The designers settled for a horizontal solution which had the optimal balance of aesthetics, cost and maintenance. The building envelope is solved with aluminum composite and rock wool in the blind parts and with high performance glass in an aluminum framed curtain wall in the spaces between pillars. Between the slat rails and the façade, a platform for external cleaning of the glass has been designed. The green façade merged with the grass around the building. This combination looks like a vibrant green-covered hill standing by the river.

材料:钢材、铝材、玻璃、复合铝材、岩棉

Materials: steel, aluminum, glass, aluminum composite, rock wool

给予绿色的关怀
GREEN CARE

 绿黄色 Green Yellow

 亮菊黄 Light Goldenrod Yellow

 暗灰色 Dim Gray

总平面图 SITE PLAN

© Hervé Abbadie

阿尔茨海默氏症患者的新住院设施,Philippon-Kalt Architects,法国,巴黎,2012年

该项目处在城市街区的中心,特征十分鲜明,在周围多座受保护的历史建筑中脱颖而出,这是由于建筑使用了Ductal®混凝土双层表皮,与附近受保护的公共花园交相呼应。外立面采用生机勃勃的绿色,搭配白色条板,与周围的树木相呼应,令人心旷神怡。

New Accommodation Facility for Alzheimer's, Philippon-Kalt Architects, Paris, France, 2012

In an environment rich in protected buildings, the project creates a very original architectural signature in the heart of an urban block thanks to its Ductal® concrete double skin, an echo of the adjacent protected open space (EVP) garden. The façade is in vigorous green paired with white slats. It responds to the surrounding trees and can make people completely relaxed and happy.

材料:Ductal®混凝土、白色条板

Materials: Ductal® concrete, white slats

中紫色 Medium Purple

 深粉红 Deep Pink

 硬木色 Burly Wood

© Grégoire Kalt

© Grégoire Kalt

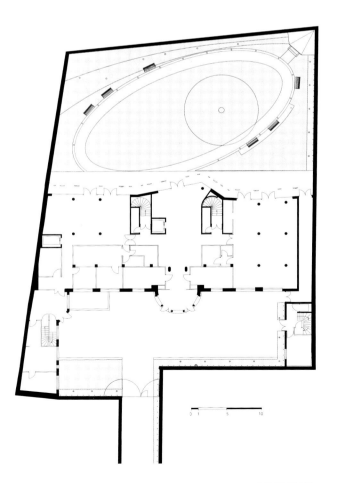
平面图 PLAN

沙棕色 Sandy Brown

 橙红色 Orange Red

 灰色 Gray

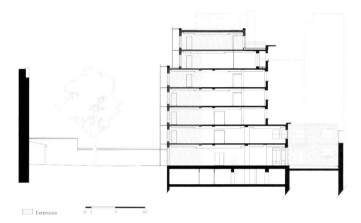

剖面图 SECTION

阿尔茨海默氏症患者的新住院设施，Philippon-Kalt Architects，法国，巴黎，2012年

为唤起患者的认知记忆、避免冷漠，各楼层均采用了鲜亮的色彩。公共起居室中悬挂着照片、海报或油画等与患者生活相关的物品，帮助他们分清各个楼层。这些视觉元素能通过刺激患者的记忆来辅助他们恢复大脑的时空经纬。

New Accommodation Facility for Alzheimer's, Philippon-Kalt Architects, Paris, France, 2012

The colors are bright and colorful on each floor to stimulate the residents' cognitive memories and avoid apathy. A photo, poster or painting that is a known reference to a period in the residents' life is in the communal living rooms to allow them to identify each floor. These visuals aid in the residents' spatio-temporal navigation by stimulating their memory.

材料：混凝土、木材、玻璃

Materials: concrete, wood, glass

© Grégoire Kalt

© Hervé Abbadie

© Hervé Abbadie

未成熟的青涩
IMMATURE YOUTH

 绿黄色 Green Yellow

 米色 Beige

 弱绿色 Pale Green

 纯黄 Yellow

 暗海洋绿 Dark Sea Green

康塞尔幼儿园，RipollTizon，西班牙，马略卡岛，2010年

　　该项目坐落在城市和乡村的交界处，位于东西轴线上的一个狭长地块中。黄绿的色调，青涩果实的颜色，像孩子们一张张天真的笑脸。幼儿园的设计需要满足两个关键性需要：第一，所有的教室和其室外私属庭院要朝向东方，以利用晨光来获取自然照明；第二，要为新建筑和原建筑之间建立衔接通道，以便该学校所有的小学生使用新的食堂，尽管新建筑要独立运营。

Consell Kindergarten, RipollTizon, Mallorca, Spain, 2010

The project is located in the edge between the urban fabric and the countryside, in a long plot oriented on an East-West axis. Yellow and green are the colors of immature fruits. They are just like the naive faces of the children. The kindergarten has to meet two key requirements: the first is that all the classrooms and its exterior private courtyards should preferably be oriented East to utilize the morning sun for natural lighting; the second is to provide an access between the new and the existing building so that the new dining room can be utilized by all the pupils of the school complex, although the new building should also operate independently.

材料：玻璃、混凝土

Materials: glass, concrete

© José Hevia

© José Hevia

© José Hevia

© José Hevia

© José Hevia

亮黄色 Light Yellow

橙色 Orange

深橙色 Dark Orange

海洋绿 Sea Green

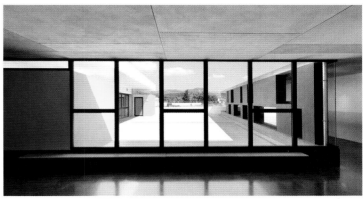
© José Hevia

© José Hevia

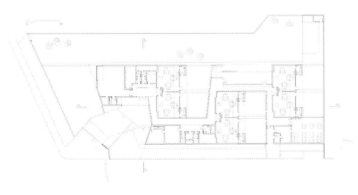
平面图 PLAN

横向剖面图 CROSS SECTION

纵向剖面图 LONGITUDINAL SECTION

康塞尔幼儿园，RIPOLLTIZON，西班牙，马略卡岛，2010年

屋顶平面通过不同的体量清晰地展现了教室和"小路"之间的关系。一方面教室保持了常规的高度；另一方面，门厅的屋顶遵循了"小路"的人字形路径：主入口处进行了抬高，迎接人们的到来，并引导着人们穿过一个更为扁平的空间进入教室，并在操场旁边形成了一个有顶的开放空间。这是一个充满活力的、可以遮阳的折叠屋顶，借用了附近橘园的颜色，为孩子们遮风挡雨，将孩子们"拥入怀抱"。

Consell Kindergarten, RIPOLLTIZON, Mallorca, Spain, 2010

The roof plan clearly shows the relation between the pavilions and the "path" using differentiated volumes. On the one hand the pavilions maintain a regular height, on the other hand the hallway roof follows the zig-zag route of the "path": gaining height to greet us in the main entrance, driving us to the classrooms through a more compressed space and finally offering us an open sheltered area next to the playground. It becomes vibrant and folded shadow roofing that covers and embraces us borrowing the colors of the orange groves nearby.

材料：玻璃、混凝土、涂料

Materials: glass, concrete, paint

一道绿色风景线
A GREEN SCENERY

 柠檬黄 Lemon Yellow

 黄绿色 Yellow Green

 闪光深绿 Lime Green

 暗绿色 Dark Green

阿布扎比洛克福特酒店，ATKINS，阿联酋，阿布扎比，2011年

　　酒店的外观就像是一个巨大的玻璃建筑。这是一座非常奢华的酒店，设计使用弯曲玻璃打造整座建筑。从该建筑的外观可以看出，整座建筑没有任何直线设计。玻璃的运用使得建筑外观呈现出蓝绿交织的迷人效果，也成为这片区域一道亮丽的风景线。

Rocco Forte Hotel Abu Dhabi, ATKINS, Abu Dhabi, UAE, 2011

The architecture is contemporary in the form of a long fluid glass structure. Its impressive elevation is a harmonious play of colored glass mosaic creating a dynamic flowing skin, combining various lively colors and tones. The mosaic effect along with the winding glass elevations break up the big building mass and create a dynamic elevation.

材料：玻璃

Material: glass

© ATKINS

© ATKINS

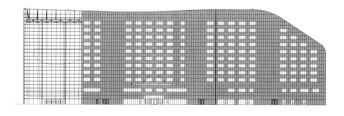

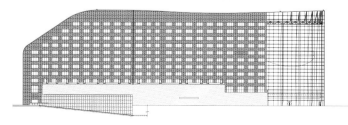

立面图 ELEVATIONS

一方洁净的空间
A PIECE OF CLEAN SPACE

橙色 Orange

橙红色 Orange Red

火砖色 Fire Brick

茶色 Tan

天蓝 Sky Blue

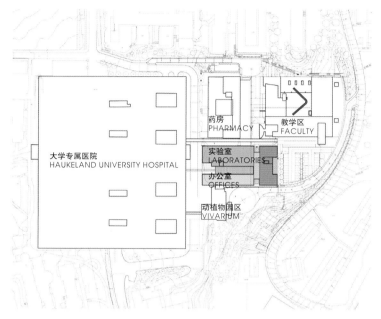

总平面图 SITE PLAN

© Jorgen True

浩克兰德大学医院实验室，C. F. Møller Architects，挪威，卑尔根，2010年

这座新建筑将研究实验室及诊断实验室汇集到一起，为医院的日常工作服务。由于建筑场地空间狭小、环境黑暗，因此建筑的顶部、立面、内墙都广泛地使用玻璃，以最大限度地增强自然光线，保证最佳的视线及开放性。

Laboratory Building at the Haukeland University Hospital, C. F. Møller Architects, Bergen, Norway, 2010

The new building brings together research laboratories and diagnostic laboratories for the daily work of the hospital. The disadvantages of this constricted, dark location led to extensive use of glass in the roof, façades and interior walls to maximize natural light, views and openness.

材料：玻璃

Material: glass

© Jorgen True

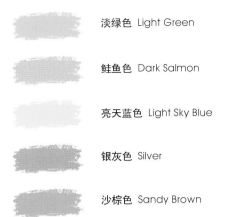

淡绿色 Light Green

鲑鱼色 Dark Salmon

亮天蓝色 Light Sky Blue

银灰色 Silver

沙棕色 Sandy Brown

© Jorgen True

浩克兰德大学医院实验室，C. F. Møller Architects，挪威，卑尔根，2010年

建筑内部有一个9层高的中庭，能够使光线深入建筑内部，有天桥横跨于办公室及实验室空间。建筑中高度灵活的实验室及相邻的工作空间在功能、技术、卫生方面达到高标准。功能的精确性和临床的洁净性也表现在精密、彩色的玻璃立面上，与周围的混凝土建筑形成一种视觉上的平衡。

Laboratory Building at the Haukeland University Hospital, C. F. Møller Architects, Bergen, Norway, 2010

A nine-story atrium brings light deep into the building, traversed by footbridges spanning between offices and laboratories. The building meets the high standards of functionality, technology and hygiene in highly flexible laboratories with directly adjacent workplaces. Functional accuracy and clinical purity is also expressed in the precise and colorful glass façades, as a visual counterbalance to the surrounding concrete buildings.

材料：玻璃

Material: glass

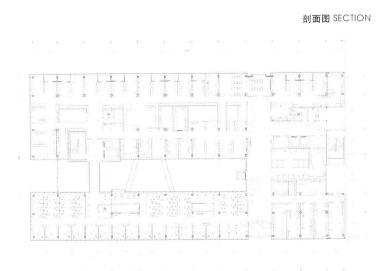

剖面图 SECTION

平面图 PLAN

雨露滋润、生机如笋
LIKE VIVIFYING BAMBOO SHOOTS BATHED IN RAIN AND DEW

 浅海洋绿 Light Sea Green

 白杏色 Blanched Almond

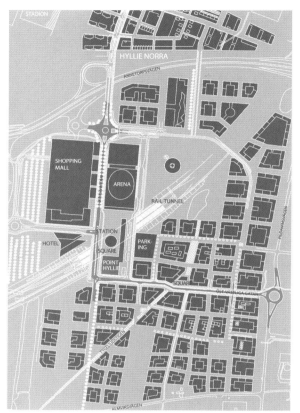

总平面图 SITE PLAN

Point Hyllie公寓楼,C. F. Møller Architects,瑞典,马尔默,2008–2013年

Point Hyllie公寓楼由同一柱体支撑底座上的4座高层建筑构成。最高的两座建筑——双子塔楼将成为瑞典的门户标志。同时,它们也是连接Hyllie住宅区和广场的纽带。建筑的设计符合人文尺度,并富有和缓的高低变化。直耸入天的建筑拥有纤长的外轮廓,而且立面向远离公共空间的方向倾斜,避免对其遮挡和影响。绿色与白色相互交叠,为建筑搭建出充满生机的外立面。高挑的玻璃窗从视觉上把建筑拉长,让建筑看上去更为高大挺拔,如雨后春笋般拔地而起。建筑利用绿色环保材料、绿色屋顶等手段,使其成为名符其实的绿色建筑和该地区一座极具特色的地标性建筑。

Point Hyllie, C. F. Møller Architects, Malmö, Sweden, 2008-2013

Point Hyllie consists of four tower blocks rising up from a column-supported base. The two tallest buildings, the twin towers, will symbolize a gateway to Sweden, but also a gateway linking the Hyllie residential area with the square. The buildings have been designed to allow the project to be adapted to human dimensions, and there is a gentle and supple transition between high and low. The tall buildings have a slender profile, and façades leaning away from the public space to avoid overshadowing and turbulence. green overlaps with white, providing a vibrant façade for the building. The slight glass windows visually lengthen the building and make it appear even taller. The building rises from the ground like the bamboo shoots after rain. By means of eco-friendly materials and green roofs, the buildings are worthy the name of green architectures and will form a distinctive landmark in the area.

材料:玻璃、钢结构、金属面板

Materials: glass, steel, metal panels

© Jorgen True

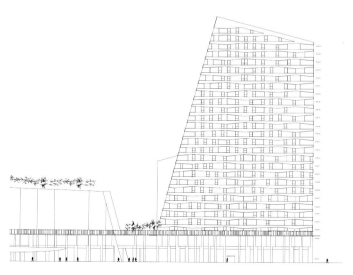

立面图 ELEVATION

梦想的舞台
STAGE FOR DREAMS

 纯黄 Yellow

 银灰色 Silver

 黛青色 Dark Cyan

 黄土赭色 Sienna

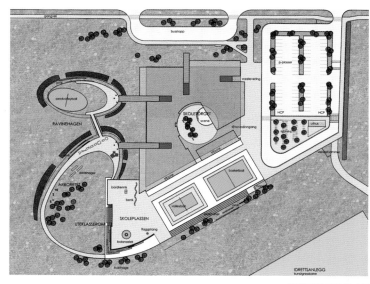

总平面图 SITE PLAN

© Rolf Estensen

© Jiri Havran

耶德鲁姆中学，Kristin Jarmund Architects，挪威，耶德鲁姆，2009年

该学校建筑设计十分紧凑，整座建筑沿着地势的起伏而建。屋顶像是从地面折叠起来的，在大片景观中创造出一个有力的、环绕的体量。在这个别致的拱形屋顶结构下方聚集了学校内所有的功能性空间。气势恢宏的开放式景观中，精致的建筑层层相连，像是房子里的房子，营造出一种丰富的内向型校园氛围。

Gjerdrum Secondary School, Kristin Jarmund Architects, Gjerdrum, Norway, 2009

The school is designed within a relatively compact building-structure. The roof is shaped as though the terrain is "folded up" from the ground, creating a powerful and encompassing dimension compared to the large surrounding landscape. The school functions are gathered under the overarching roof-structure. An open landscape made up by smaller buildings, as "houses within the house", creates a rich inner atmosphere.

浴火卫士
FIRE GUARD

 火砖色 Fire Brick

 深橙色 Dark Orange

 橙色 Orange

 纯黄 Yellow

 硬木色 Burly Wood

 绿黄色 Green Yellow

 银灰色 Silver

加瓦消防站，Mestura Arquitectes，西班牙，巴塞罗那，加瓦，2010年

该建筑共两层，呈立方形状。建筑的平面十分紧凑，设计体现了其功能性，确保在发生火灾和突发事件时能够快速行动。室内大胆地运用了红绿对比色，红色依绿色而明媚，绿色依红色而浓郁。色彩鲜明亮丽，却不冲突，有很强的视觉冲击力，与消防站的功能性贴切。

Fire Station in Gavà, Mestura Arquitectes, Gavà, Barcelona, Spain, 2010

This two-floor building is designed with a cubic shape. The floor plans are very compact and functionally designed to support speedy action in case of fire and emergencies. The interior uses the contrast colors of red and green, which complement and reinforce each other. The colors are both bright, with no clash, creating a strong visual impact, which is very suitable for the fire station theme.

材料：钢筋混凝土、玻璃

Materials: reinforced concrete, glass

东北立面图 NORTHEAST ELEVATION

西北立面图 NORTHWEST ELEVATION

© Pedro Pegenaute

© Pedro Pegenaute

© Pedro Pegenaute

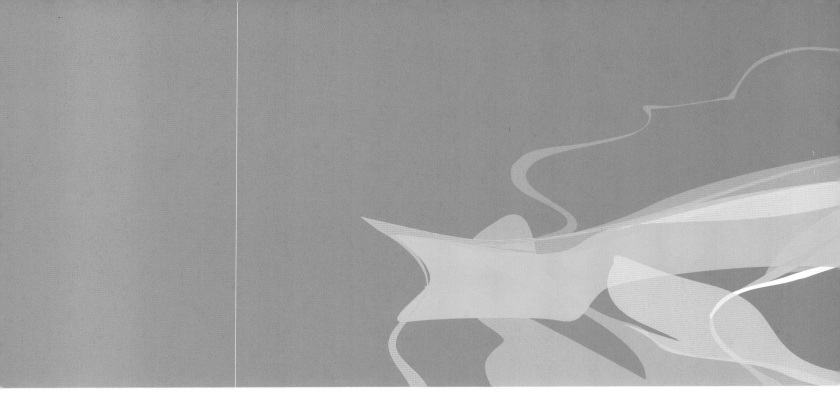

■紫，帛青赤色。——《说文》■梦幻，亦梦亦幻，宛若仙境一般的唯美空间，给人们呈现出一种近乎不可思议的超凡感受。■梦幻的代表色为紫色，象征高贵、灵感、神秘、浪漫。■紫色是一种红色和蓝色的混合，虽然这两种颜色在心里和生理上的效果正好相反。所以，浅紫色代表生活中的明亮面，深紫色代表黑暗的力量。■偏红的粉紫色适于打造温婉柔和的环境；偏蓝偏冷的紫色能放松神经，有助于营造如梦如幻的感觉。■营造梦幻的效果不仅要用色，还要同时把控光、形、质各方面协和运作。可以说，梦幻是建筑设计灵感的升华，是用空间用色的最高境界。■总的来说，紫色用于墙壁、顶棚、地板可产生多种效果，也是灯光最常使用的颜色。

■ Purple is made by mixing blue and red together. ■ Fantasy is a combination of dreams and imagination. It can create fairyland-like aesthetic spaces and brings people an incredible and extraordinary feeling. ■ The representative color of fantasy is purple, which stands for nobleness, inspiration, mystery and romance. ■ Purple mixed by red and blue, though the two colors have quite the opposite effects on people's mind and body. Therefore, the lighter purple represents the brightness of life, while the darker purple represents the power of the dark side. ■ Pinkish purple is suitable to create a mild and soft environment. While cool bluish purple can promote the sense of depth, relax people's nerves and help to create a dreamlike ambience. ■ Color is not the only element for creating dreamlike effects, the coordination and management of light, shape, material is also necessary. It can be argued that fantasy is the distillated inspiration in architecture design and the highest level of color schemes for spaces. ■ All in all, purple can make various effects when applied on walls, ceilings and floors, and it is most commonly used for lighting.

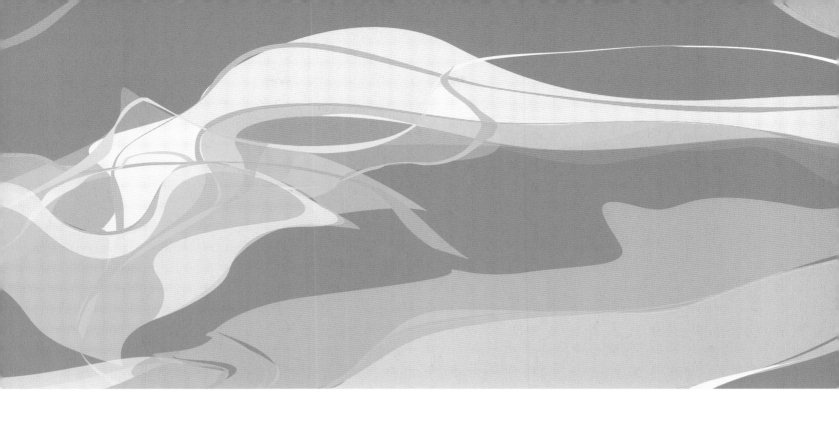

紫之梦幻
PURPLE
157

穿越时空的水晶球
A CRYSTAL BALL ACROSS TIME AND SPACE

 淡紫色 Lavender

 中紫色 Medium Purple

 紫兰色 Indigo

 矢车菊蓝 Cornflower Blue

 番茄红 Tomato

Flockr临时展馆，SO-IL，中国，北京，2010年

　　单色的材质如何营造缤纷绚丽的效果？用镜面材质反射周遭的景色可谓最讨巧的一种。SO-IL即用数以千计的淡紫色镜面金属板铺就，弧面亮紫色的镜面将周围的楼宇、街道、甚至行人通通解构、扭曲、再呈现，营造出一种超现实如梦如幻的奇异世界。SO-IL的缤纷效果，源自周围绚丽的环境色，而自身的紫色为水滴般的建筑带来时尚感，仿佛一个可以穿越时空的外来建筑。

Flockr Pavilon, SO-IL, Beijing, China, 2010

 How to create a gorgeous effect with single-colored materials? The most subtle way is to use mirrored materials to reflect the surrounding landscapes. SO-IL is covered with thousands of tinted mirrored panels. The bright purple cambered surface reconstructs, distorts and re-presents the surrounding buildings, streets, even people, creating a dreamlike surreal world. The colorful effect of SO-IL is originated from the color of the surrounding. Purple adds a modern sense to the water drop-like building, as if the building came from outer world and could travel across time and space.

材料：有色镜面金属板、钢结构

Materials: tinted mirrored panels, steel structure

© Iwan Baan

立面图 ELEVATION

© Iwan Baan

宝石与梦境的圆舞曲
THE WALTZ OF GEMSTONES AND DREAMS

淡青色 Light Cyan

烟灰色 Smoke Gray

纳瓦白 Navajo White

桃肉色 Peach Puff

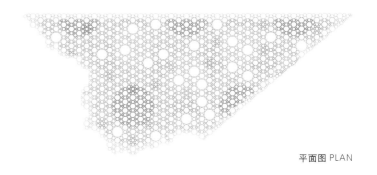

平面图 PLAN

宝格丽展亭，NaNA，阿联酋，阿布扎比，2012年

展亭由2 272根不同长度和直径的有机玻璃圆管通过大约10 000个连接件组成。对这种光洁材料进行"切"与"割"，加上与环境的对比，突显出宝格丽这个意大利名贵奢侈珠宝品牌的卓越与典雅。整个展亭就像一片轻柔的透明薄纱，像一颗宝石般闪闪发光。白天，利用淡淡的橙色光晕从内部晕染开来，使"宝石"变得温软、莹润。

Bulgari Pavilion, NaNA, Abu Dhabi, UAE, 2012

The pavilion consists of 2,272 acrylic tubes of different lengths and diameters with approximately 10,000 connections between them. The bright materials, through cutting, contrast with the environmental material and emphasize the prominence and grace of the Bulgari, the Italian luxury brand. The entire pavilion is just like a soft, light, transparent veil or a shining gemstone. During the day the faint orange light glows from the interior, making the gemstone warm, soft and sparkling.

材料：亚克力管

Material: acrylic tubes

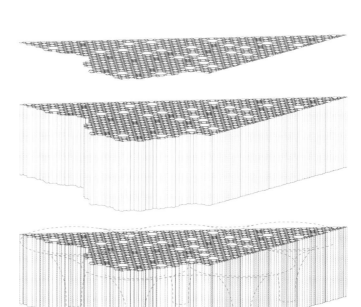

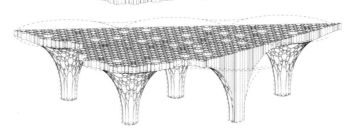

概念图 FINAL ARRANGEMENT CONCEPT

© NaNA

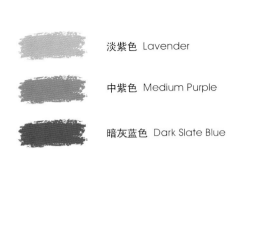

淡紫色 Lavender

中紫色 Medium Purple

暗灰蓝色 Dark Slate Blue

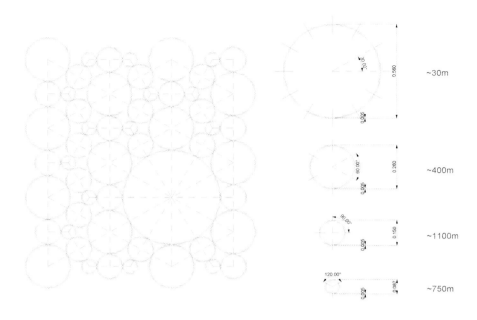

~30m

~400m

~1100m

~750m

装配模式 ASSEMBLY PATTERN

© NaNA

浅粉红 Light Pink

 玫瑰棕色 Rosy Brown

 沙棕色 Sandy Brown

© NaNA

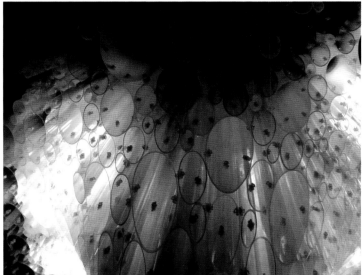

© NaNA

宝格丽展亭，NaNA，阿联酋，阿布扎比，2012年

展亭像一颗粗糙的宝石，设计概念就是切削，然后打磨抛光，形成新的空间与环境。夜晚，玫瑰色与淡紫色的灯光交相呼应，把空间渲染得无比梦幻与神秘。清幽的色彩充斥在空间的每一个角落，令人们仿佛置身于用宝石编织的梦境之中，奇妙而绚丽。身处这样一个梦幻的空间之中，人们相互倾听、诉说，连心情都会变得轻松愉悦起来。

Bulgari Pavilion, NaNA, Abu Dhabi, UAE, 2012

Much like the rough gemstones, the design concept was to slice and cut, smooth and polish the form to result in new spaces and environmental conditions. In the night, the rose and lavender lights add radiance and beauty to each other, rendering the space much fantastic and mysterious. The secluded colors spread to every corner of the space, seeming to have brought people into a fancy and colorful dreamland which is composed of gemstones. Listening to and talking with each other in such a space, people's mood will surely perk up.

材料：亚克力管

Material: acrylic tubes

© NaNA

紫之梦幻 | PURPLE

炫彩万花筒般的奇幻世界
A KALEIDOSCOPIC WORLD

爱丽丝蓝 Alice Blue

淡紫色 Lavender

桃肉色 Peach Puff

暗淡灰 Dim Gray

深天蓝 Deep Sky Blue

纯黑 Black

中川化工CS设计中心（"万花筒"展览），Emmanuelle Moureaux Architecture + Design，日本，东京，2009年

设计师为中川化工CS设计中心设计了办公室和陈列室。灵感来源于万花筒，设计师使用超过1 100种颜色，创造出惊人的办公空间。展览每次仅侧重于一种色系——黄色、红色、绿色、蓝色或黑色。每个月的同一时间，空间设计变换不同的颜色，并且强调色调的变化，就像走进了一个缤纷万花筒的世界。展览旨在重新发现普通平凡颜色中的美，通过多层玻璃面板的精心分层，使不同色块之间也相互重叠，产生出万花筒般的神奇效果。

The Nakagawa Chemical CS Design Center, "Kaleidoscope" Exhibition, Emmanuelle Moureaux Architecture + Design, Tokyo, Japan, 2009

The designer designed the offices and showrooms for Nakagawa Chemical CS Design Center. Taking inspiration from a kaleidoscope, the designer used over 1,100 colors to create the stunning office spaces. The "kaleidoscope" exhibition focused on one color at a time such as yellow, red, green, blue or black. At the same time every month, the space displayed a different color, changing the hues like a kaleidoscope world- the exhibition aimed to rediscover the beauty of ordinary colors. The glass panels were carefully layered and the same color blocks overlapped each other, creating a wonderful kaleidoscope effect.

材料：玻璃、涂层

Materials: glass, coating

© Hidehiko Nagaishi

© Hidehiko Nagaishi

© Hidehiko Nagaishi

© Hidehiko Nagaishi

爱丽丝蓝 Alice Blue

 森林绿 Forest Green

 纯黄 Yellow

 猩红 Crimson

 灰菊黄 Pale Goldenrod

中川化工CS设计中心（"万花筒"展览），Emmanuelle Moureaux Architecture + Design，日本，东京，2009年

方形彩块在同一色系的不同色调中变换，由近到远，乃至消失不见……使空间产生出一种无尽头的虚无感。利用不同色系对人心理上的影响，使空间所产生出的意义也相对改变。蓝色是最为清冷的颜色，能让空间的气氛安静了下来，一切喧嚣似乎都被隔绝在屋外；黑色有时会使人们感受到那种令人难以言状的共性，带有神秘的色彩；绿色象征着生命、平衡、和平和生命力，令办公的环境生机勃勃，春意盎然；黄色是一种能使人引起愉快遐想的颜色，它渗透出来的灵感和生气使人欢乐和振奋，使空间温暖起来；红色给人以热烈、兴奋之感，使环境充满激情。

The Nakagawa Chemical CS Design Center, "kaleidoscope" exhibition, Emmanuelle Moureaux Architecture + Design, Tokyo, Japan, 2009

The square panels of the same color system change their hues from near to far and finally disappear from sight…which creates a create a sense of nobility. Making use of the influences various colors have on people's psychology, the designer endowed the space with different meanings. Blue is the coolest color which can calm down the atmosphere. All the clamors seem to have been kept outside; black is mysterious and can sometimes make people feel the common characteristics; green stands for life, balance, peace and vitality. It can fill the office environment with vigor and spring breath; yellow is a color that can remind people of pleasant experiences. The inspiration and vigor it infiltrates can please people and warm the room; red are associated with enthusiasm and excitement and can fill the environment with passion.

材料：玻璃、涂层

Materials: glass, coating

© Hidehiko Nagaishi

© Hidehiko Nagaishi

© Hidehiko Nagaishi

紫之梦幻 | PURPLE

山脚下的梦想城堡
THE DREAM CASTLE AT THE FOOT OF THE MOUNTAIN

 烟灰色 Smoke Gray

 亮钢蓝 Light Steel Blue

 灰石色 Slate Gray

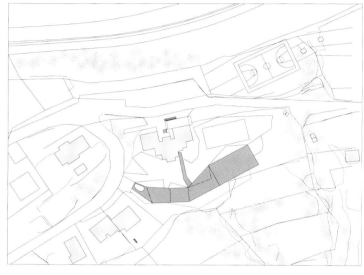

总平面图 SITE PLAN

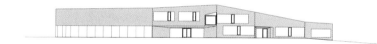

立面图 FAÇADE

波维尼尔小学扩建，Bonnard Woeffray Architectes，瑞士，波维尼尔，2010年

小学的扩建部分就像是穿上了一件巨大的防护盔甲一样，形成一道抵御雪崩的天然屏障。唯一设有紧急出口的立面覆有金属外皮，镜子般的等高玻璃从中穿过，但丝毫没有影响其连贯性。其他立面均种植了绿化植被。原始铝材的光泽反射着对面山坡的光线，使建筑立面的色彩因环境光线的强弱而变化，为建筑场地增添了新的维度。

Bovernier Primary School Extension, Bonnard Woeffray Architectes, Bovernier, Switzerland, 2010

Like an immense suit of protective armor, the extension project forms a natural shield against avalanches. The color scheme, here pastel, there vivid, bathes the north-facing spaces in a range of hues. The sole emergent façade is clad in a metallic skin pierced with mirrored, flush glazing that does not impinge on the continuity. The other façades are greened with vegetation. The luster of the raw aluminum reflects the light of the hillside opposite, giving the site a new dimension.

© Hannes Henz

材料：铝材、混凝土、玻璃

Materials: aluminum, concrete, glass

金菊黄 Goldenrod

 巧克力色 Chocolate

 橙红色 Orange Red

 浅粉红 Light Pink

© Hannes Henz

波维尼尔小学的扩建，Bonnard Woeffray Architectes，瑞士，波维尼尔，2010年

　　粉红、紫、黄等缤纷的色彩充斥在室内的每一个角落，像是孩子们那些童真美好的梦想，会在这里一一实现。内部空间采取了并行排列的布局，既能够相互连接又可独立运行。这些功能区的排列像一列火车一样，遵循了与村庄和学校运作相关的逻辑顺序：首先是日间幼儿园，中间是小学，最后是健身房。扩建部分通过悬于操场上方的一条封闭走廊与学校原建筑相连。

Bovernier Primary School Extension, Bonnard Woeffray Architectes, Bovernier, Switzerland, 2010

Different from the façade, bright colors such as pink, purple, yellow occupy every corner of the interior spaces. These cute colors are like sweet childhood dreams which will come true right here. The spaces are juxtaposed, both linkable and separable. Rather like a train, these functions follow a logical sequence relative to the village and the functioning of the school: first comes the day nursery, followed by the primary school in the centre and the gym at the end. The enlargement "plugs" into the existing school via a covered walkway flown above the playground.

材料：混凝土、玻璃

Materials: concrete, glass

© Hannes Henz

银灰色 Silver

 纯黄 Yellow

 暗兰花紫 Dark Orchid

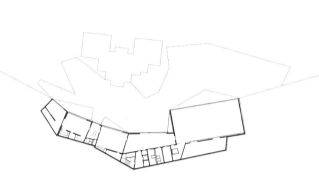

一层平面图 GROUND FLOOR PLAN

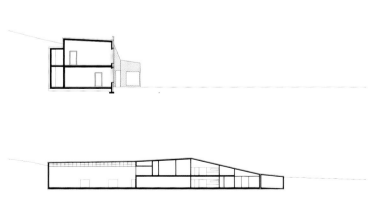

剖面图 SECTIONS

紫之梦幻 | PURPLE | 171

历史的展现、唯美的重生
REAPPEARANCE OF HISTORY, REBIRTH OF AESTHETICISM

灰菊黄 Pale Goldenrod

亮菊黄 Light Goldenrod Yellow

橄榄 Olive

钢蓝 Steel Blue

午夜蓝 Midnight Blue

1 大厅
2 中庭走廊
3 画廊
4 登记/艺术处理
5 服务/支援
6 装卸码头
7 安全防护

1 Lobby
2 Atrium Gallery
3 Gallery
4 Registrar / Art Handling
5 Service / Support
6 Loading Dock
7 Security

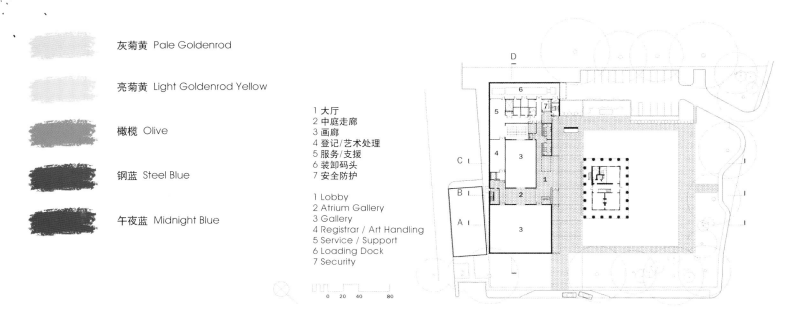

一层平面图 FIRST FLOOR PLAN

© Timothy Hursley

保罗和露露希利亚德大学艺术博物馆，Eskew+Dumez+Ripple，美国，路易斯安那州，拉斐特，2004年

该博物馆分为古建与现代建筑两部分。建筑师将旧建筑翻新，全部刷白，使其显得更加大气端庄，蕴涵文化气息。新建筑部分是钢结构构筑物，其中三面是预制混凝土墙，一面是玻璃幕墙，透过玻璃幕墙的侧面是旧博物馆。当游客观望旧建筑时，他们可以看到其建筑结构似乎比新建筑的结构要沉稳得多，并且旧建筑"深深地植根于土地"之中，新建筑则几乎是漂浮于地面的。最为经典的是PHA工作室设计的灯光方案，这个方案需要在楼板中开孔，将电子输送霓虹的电子软管安装于顶棚之中。每天晚上，深蓝色的灯光都会笼罩着整个建筑，使得它更加优雅神秘。而原始建筑则采用了一个传统的色调，简洁，清爽。

Paul and Lulu Hilliard University Art Museum, Eskew+Dumez+Ripple, Lafayette, Louisiana, USA, 2004

The museum consists of an old apart and a modern building. The architects renewed the old building and painted it all white to create a grand, elegant and cultural building. The new building is a steel-frame structure with precast-concrete walls on three sides and a glass curtain wall on the fourth side facing the old museum. When visitors look back at the old building they see a structure that appears heavier than the new one and "rooted to the ground" while the new building appears to almost levitate. Working with most classical PHA Lighting Design, the architects created a lighting scheme that cast the building, every evening, in a deep blue light coming from cold cathode tubes set within a perforated metal ceiling. The building looks more graceful and mysterious with the lighting. In contrast, they cast the exterior of the original building in a traditional, clean, white light.

© Timothy Hursley

材料：钢结构、混凝土、玻璃

Materials: steel structure, concrete, glass

水晶花瓣
CRYSTAL PETALS

　　　　雪白色 Snow

　　亮天蓝色 Light Sky Blue

　　暗黄褐色 Dark Khaki

"瓣"亭，Orproject，中国，北京，2012年

　　"瓣"是在2012年北京设计周时建立的。其中文名称含义为花瓣，形状就像由弯曲的花瓣构成的花朵。"瓣"是一个由卷曲的高分子聚合物板构成的自承重结构体，并用一片片"花瓣"搭建出了形状和体量。基于Orproject对薄板的各向异性形态学的研究，亭子的几何造型采用了正交结构，用单弧度元素创造出了一系列的支撑柱体、弧形和拱形。亭子的线条引领着观者的视线穿过结构体，看向天空。"瓣"就像一个巨大的花朵一样，停悬在北京老胡同的屋顶上空。

"Ban" Pavilion, Orproject, Beijing, China, 2012

Ban is a pavilion which has been constructed for the Beijing Design Week 2012. The Chinese title refers to flower petals, and similar to the way that the shape of a flower is created by its bent petals. Ban is constructed from bent polymer sheets which form a self-supporting structure and create shapes and volume from a multitude of leaves. Based on Orproject's research into anisotropic sheet morphologies, the geometries have here been used in a structural orthogonal orientation and form a system of columns, arches and vaults, all based on single-curved elements. The resulting field of lines takes the viewers' eye across the structure and into the sky, and like a giant flower Ban is hovering in the air above Beijing's ancient Hutong roofs.

材料：聚合物板

Material: polymer sheets

© Jasper James & Orproject

© Jasper James & Orproject

通往星空的时光隧道
THE TIME TUNNEL TO THE STARRY SKY

 猩红 Crimson

 紫色 Purple

 紫兰色 Indigo

 午夜蓝 Midnight Blue

 闪兰色 Dodger Blue

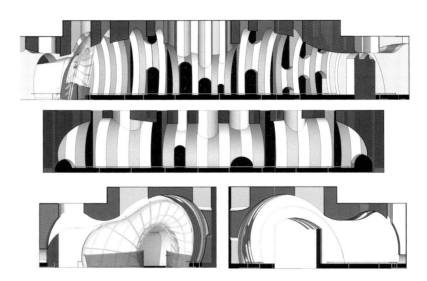

剖面图 SECTIONS

© Hedrich Blessing Photographers

© Hedrich Blessing Photographers

阿德勒天文馆内的克拉克家族欢迎画廊，Thomas Roszak Architecture，美国，芝加哥，2011年

通过向纺织行业的专家求教，以及多种方案的甄选，Roszak设计团队发现了一种线条精密清新、外观天衣无缝的织物。通过间断地对织物进行分层，沿参观通道形成深浅不一的薄片，象征着外太空的"时间切片"。该织物拥有光滑的质感，纯白的色调，无论是单层还是双层使用，透明度都恰到好处。蓝红相间的灯光设计，透过织物薄片晕染开来，为参观通道笼罩上一层似太空般神秘虚幻的氛围。

Clark Family Welcome Gallery at the Adler Planetarium, Thomas Roszak Architecture, Chicago, USA, 2011

Roszak's team consulted "fabricologists", experts in the fabric industry, sorting through thousands of choices before discovering the perfect options. The team didn't settle until they found a fabric with precise, clean lines embodying the desire for a seamless look. The team layered fabric in alternating areas creating darker and lighter "slices" along the visitor's pathway with each section, representing a "time slice" in outer space. It is a smooth textured, pure white fabric with just the right amount of transparency used both in single and double layers. The blue-red lighting permeates through the "slices" and creates an ambience as mysterious as the outer space.

材料：铝、织物

Materials: aluminum, fabric

© Hedrich Blessing Photographers

熏衣草淡紫 Lavender

 石蓝色 Slate Blue

 暗灰蓝色 Dark Slate Blue

© Hedrich Blessing Photographers

© Hedrich Blessing Photographers

阿德勒天文馆内的克拉克家族欢迎画廊，Thomas Roszak Architecture，美国，芝加哥，2011年

侧立面外形如同瀑布飞流而下，如此形成的道道坡面也成为设计元素，象征随着时间的推移记录天文事件。设计师使用铝管和涤纶织物建造墙壁，创造出一个超凡脱俗的空间。最重要的是，这种织物不但可以反射空间的LED照明系统，还能吸收投射在织物墙面上的多媒体视频光线。画廊经过精心布局，参观者在画廊中移动位置，对空间的感知就会如空洞一般分开、聚合、扩大、缩小，不断发生变化。通过与视频、动画、声音、灯光方面专家的广泛合作，使互动式展品具有层叠投影图像、运动探测光线和声音效果。

Clark Family Welcome Gallery at the Adler Planetarium, Thomas Roszak Architecture, Chicago, USA, 2011

Movement through the gallery is delineated by cascading sheer fabric planes, which represent astronomical events over time. Using aluminum tubing and polyester fabric constructed walls provided a dramatic design for the creation of space that feels other-worldly. Most importantly, the fabric had to have the ability to reflect the space's LED lighting system, but also absorb light where mixed-media video is projected on the fabric walls. The floor-plan is laid out such that by shifting one's position in the gallery, the perception of space changes as the voids separate, join, expand, and contract. Extensive collaboration was done with educators and experts in video, animation, sound, and lighting to create inspired interactive exhibits with layered projected images, motion-detecting light, and sound effects.

材料：铝、织物

Materials: aluminum, fabric

城海中的小憩与欢愉之岛
A PLEASURE ISLAND IN THE CITY

 熏衣草淡紫 Lavender

 浅粉红 Light Pink

 中紫色 Medium Purple

 雪白色 Snow

浅玫瑰色 Misty Rose

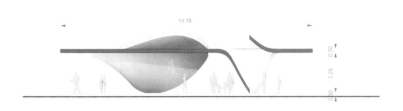

剖面图 SECTIONS

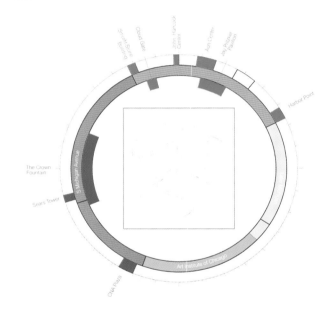

分析图 ANALYSIS DIAGRAM

伯纳姆展馆，UNStudio，美国，芝加哥，2009年

　　UNStudio设计的这座装置展馆深具雕塑感，具有高通达性，是都市生活的活化剂。装置展馆以直线和斜线的空间均衡为基础，展现了360°全方位的景致。受伯纳姆分层构造的启发，展亭在地板、墙壁和顶棚之间设置了一定的倾斜度，打造出一个浮动的多向性空间。横向和纵向平面的分层设计转化为对空间连续、可变而流畅的诠释。白天，洁净的纯白色立面与公园内的高大绿植融为一体，形成一幅安宁祥和的画面，为周围喧嚣的都市带来一处宁静，行人可在此小憩；夜晚，各色灯光渲染了展馆的白色立面，蓝、红、紫……使展馆在暗夜里格外引人瞩目。行人被绚丽的色彩吸引，欢聚于此，热闹非凡。

Burnham Pavilion, UNStudio, Chicago, USA, 2009

The UNStudio pavilion is sculptural, highly accessible and functions as an urban activator. The pavilion is based on a similar spatial confrontation of the orthogonal and the diagonal, but extends the vistas to encompass the entire spectrum of 360 degrees. Inspired by Burnham's tectonic layering of the city, the pavilion introduces a gradient between floor, wall and ceiling, introducing a floating and multidirectional space. The hierarchy of the horizontal or vertical plane is converted into an understanding of a space as continuous, transformative and fluid. During the day, its clear white façade merge with the tall greenery in the park, presenting a peaceful picture. It provides a silent shelter for visitors to take a break. In the night, with blue, red and purple light covering the white façade, it looks extremely charming in the dark. Visitors are attracted by the colorful light and gather around the pavilion.

© Christian Richters

材料：钢、胶合板

Materials: steel, plywood

© Christian Richters

© Christian Richters

霓虹光晕
THE HALO OF NEON LIGHTS

 银灰色 Silver

 闪兰色 Dodger Blue

 中暗蓝色 Medium Slate Blue

 熏衣草淡紫 Lavender

 重褐色 Saddle Brown

克林锦湖文化综合大厦,UnSangDong Architects,韩国,首尔,2010年

  建筑师旨在建造一座建筑,为公司和消费者建立标识以促进相互交流,并赋予建筑独特的身份。为了建立客户的品牌标识性,建筑师在这座文化综合体的设计中采用了"品牌的都市雕塑"理念。"锦湖"的商标图案通过建筑形体被转化为品牌标识,初衷是强调和谐。通过建筑的演绎,和谐的理念通过起伏的轮廓与城市和社会连为一体,并得到了进一步的突显和深化。建筑师在设计中和谐地揉入了自然、生活和城市的诸多元素以及和谐的精髓,使之与城市相呼应,并在建筑中幻化成回声波和波浪的图案,这是该项目的品牌化设计策略。

Kring Kumho Culture Complex, UnSangDong Architects, Seoul, South Korea, 2010

  The architects are intended to construct an architectural building to create codes for companies and consumers to facilitate communication, and then embed an identity in that architecture. "Urban Sculpture for Branding" is the concept that they applied when designing this culture complex to realize brand identity for the client. The brand image of Kumho was transformed as brand identity through the architectural shaping process. The starting point of Kumho brand was putting emphasis on "harmony". Through the architectural interpretation process, the phenomenon of harmony was connected and emphasized to the city and society with undulation, then the notion of undulation was deepened. The architects wanted to gather various elements of nature, life, and city harmoniously and that essence of harmony to rush out to the city creating echo and undulations, and that became the brand scenario of the project.

材料:不锈钢、玻璃、LED灯

Materials: stainless steel, glass, LED lights

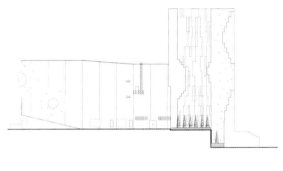

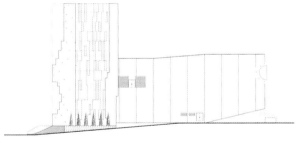

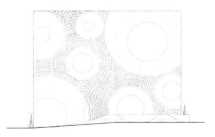

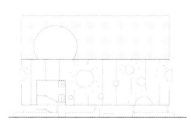

立面图 FAÇADES

© Sergio Pirrone

© Sergio Pirrone

© Sergio Pirrone

	雪白色 Snow
	深天蓝 Deep Sky Blue
	亮天蓝色 Light Sky Blue
	熏衣草淡紫 Lavender
	中紫色 Medium Purple
	银灰色 Silver

© Sergio Pirrone

克林锦湖文化综合大厦，UnSangDong Architects，韩国，首尔，2010 年

在城市中突显自身形象，并同时吸收城市的能量，这是这座大型"回声容器"诞生的初衷，它的出现还衍生出了"梦想"的理念。建筑师旨在将其打造成一座日夜屹立于城市中的丰碑，尤其是使之成为一座在黑夜中熠熠闪光的雕塑。此外，建筑的内部空间为纯白色，容纳了各种文化项目。纯白的空间实现了空间的超现实主义，宛若一座舞台一般。舞会会根据演出类型和戏剧情节而改变布景，与之类似，该建筑内部的复合空间能够灵活调整以满足各种需求。

Kring Kumho Culture Complex, UnSangDong Architects, Seoul, South Korea, 2010

Projecting images to the city and sucking up the energy of the city at the same time, gigantic "container for echo", was born that way, which leads to the concept of "Dream" as well. We wanted this to be the monument of the city day and night, specifically, a lighting sculpture when it is dark. Moreover, it becomes a pure white space when entered into, and contains all and any types of cultural program. The pure white space which achieves spatial surrealism is acting as a stage for performance. Just like a stage changes itself to conform to the types and story of a play, this becomes a compound space to fit itself to diverse needs.

材料：不锈钢、玻璃、LED灯

Materials: stainless steel, glass, LED lights

© Sergio Pirrone

柔光萦绕、花瓣纷飞
SOFT LIGHT HOVERING FLYING PETALS

 熏衣草淡紫 Lavender

 亮菊黄 Light Goldenrod Yellow

 暗灰蓝色 Dark Slate Blue

 闪兰色 Dodger Blue

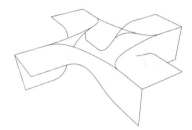

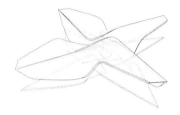
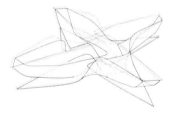

几何原理分析图 GEOMETRIC PRINCIPLES DIAGRAM

© James D'Addioh

© James D'Addioh

新阿姆斯特丹广场与展亭，UNStudio，美国，纽约，2011年

　　从曼哈顿区的摩天大楼上俯瞰，展亭无论在白天还是夜晚都清晰可见。当夜晚降临时，暖黄色的灯光温馨淡雅，萦绕在每个功能区域中，给人一种家的温暖。台面下设置了不同颜色的光源，使区域更具识别性。样式上的重复性是动感而非静态的，衍生出了多变的视角和景致。展亭的几何形体将其设计意向表现得淋漓尽致。4座花瓣形侧翼与场地的不同定位和主体建筑的功能多样性相呼应。流畅的几何形体模糊了每座侧翼的内部和外部的界限。弯曲的弧面使顶棚、墙壁和地板难分难辨，使建筑形体与周边公园的景致相得益彰。

New Amsterdam Plein & Pavilion, UNStudio, New York, USA, 2011

 The geometry of the Pavilion fully expresses its programmatic intentions. The attendant "flowering" or opening of the four wings of the Pavilion responds to varying orientations on the site as well as variety within the main program. The geometric loop is introduced to virtually obscure the boundary between the ceiling, wall and floor and to promote integration of the built form with the surrounding park. The Pavilion will be readily visible both day and night from the surrounding skyscrapers of Lower Manhattan. Repetition is dynamic rather than static, allowing varying viewpoints and perspectives to be created. When night falls, the yellow light hovers within each functional zone and therefore brings people sweet, graceful family warmth. The light sources of different colors set below make the zones more recognizable.

材料：Corian®抛光表面、烧结玻璃外壳、抛光铝格栅

Materials: Polished Corian® surfaces, fritted glass enclosures, polished aluminum grill

孤静相随、归之净土
THE SOLITARY LAND TO RETURN TO

 弱绿宝石 Pale Turquoise

 闪兰色 Dodger Blue

 灰菊黄 Pale Goldenrod

概念图 CONCEPT DIAGRAMS

更衣室，UNStudio，意大利，威尼斯，威尼斯建筑双年展，2008年

  这一装置采用了蜿蜒曲折的轨迹，在各种投影和灯光效果的映照下显得错综迷离。这些瞬变的效果不仅使空间玄幻多变，也同样把色彩和光线映射到了游客的脸庞。蓝色是永恒的象征，是色彩中最冷的颜色，是宁静，是忧郁……能让人更加清醒，去思考，去面对。淡紫色又从另一个角落渗透开来，给予一丝温暖与幻想。别墅样式的装置设计充满想象力，为打造私密空间创造了条件。从巨大的窗户俯瞰，可以看到闪亮的表面，光秃的转角，挑高的空间，粗糙的墙壁等。威尼斯双年展是建筑作为一种空间艺术形式回归其核心本质的最后净土之一。在这里，建筑本身就是"现代世界中的'家'"。

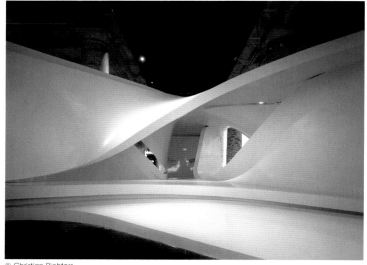
© Christian Richters

© Christian Richters

The Changing Room, UNStudio, Venice Biennale of Architecture, Venice, Italy, 2008

 The installation entails a curved trajectory which is made complex by different projections and light effects. These transient effects make the room change, but they also alter the appearance of the visitor who is showered with color and light. Blue, the coolest color, stands for eternity, silence and melancholy and so make people soberly ponder and face all. Lavender infiltrates from another corner, offering some warmth and fantasies. The imaginative installation, in simulacra of contemporary villa settings, creates conditions for producing private spaces. The large window overlooking the shimmering surface, the stark corner, the tall space, the raw wall and so on. The Venice Biennale is one of the last places where architecture returns to its core identity as a spatial art form, where architecture itself can be "at home in the modern world".

© Christian Richters

功能与色彩的艺术之舞
THE ARTISTIC DANCE OF FUNCTION AND COLOR

亮菊黄 Light Goldenrod Yellow

灰色 Gray

黄绿色 Yellow Green

纯蓝 Blue

宝石碧绿 Aquamarine

纯黄 Yellow

猩红 Crimson

1号停车场，Elliott + Associates Architects，美国，俄克拉何马州，俄克拉何马城，2011年

"停车作为艺术"是设计师形容他们在俄克拉何马为Chesapeake Energy设计的Car Park One停车楼。建筑的外立面由交错的金属面板构成，白色的立面使观赏者忽略色彩，更多地关注结构与功能上的设计理念。中庭的人行轴线将人们引向场地，各楼层设有缤纷的灯光。彩色的灯光随错列的面板渗溢氤氲，令人一见难忘。室内以灯光和色彩的加入而成为一处美妙之地。为了凸显色彩和提供照明，建筑师在每条横梁上首尾相连地安装了T8线型荧光灯，以照亮上方的顶棚。红色、黄色、绿色、蓝色、白色——停车场的每一层都有一个主色调，绚烂的色彩在美化建筑的同时，还具有强烈的标志性，使楼层之间更容易区分。功能与色彩两者完美结合，色彩不再是单纯的装饰，而且具备了功能性。

Car Park One, Elliott + Associates Architects, Oklahoma City, Oklahoma, USA, 2011

"Parking as art" is how the architects describe Car Park One they designed for Chesapeake Energy in Oklahoma. The exterior of the building features staggered metal panels. The white façade makes visitors to neglect the color and turn to its design concept of structure and function. Atrium provides a pedestrian axis to direct people to the campus with colored lighting coordinated at each level. Colored lights provide a memorable effect, as light bleeds through staggered panels. Light and color are introduced into the structure to make the building a destination. To accentuate colors and provide illumination, the architects installed T8 linear fluorescent lights end-to-end on each cross beam to uplight the ceiling plane. Each level of the car park features a primary color – Red, Yellow, Green, Blue and White – not only to beautify the facility, but aslo for users to easily navigate through the facility. Function and colors are perfectly combined. Colors serve no longer as decoration but become functional.

材料：金属面板、彩灯

Materials: metal panels, colored lighting

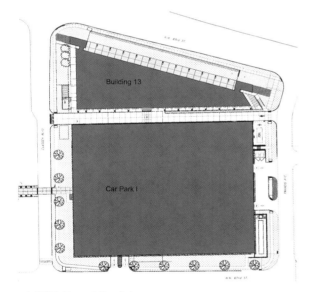

1 汽车入口	1 Car Entrance
2 汽车出口	2 Car Exit
3 服务驱动	3 Service Drive
4 桥	4 Bridge

总平面图 SITE PLAN

© Scott McDonald

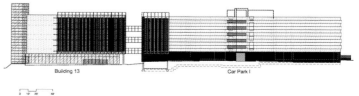
立面图 ELEVATION

舞动色彩、赋予意义
DANCING COLORS, GIVE MEANING

 深灰色 Dark Gray

 棕色 Brown

 亮黄色 Light Yellow

 灰色 Gray

 鹿皮色 Moccasin

 淡绿色 Light Green

 闪兰色 Dodger Blue

 纯红 Red

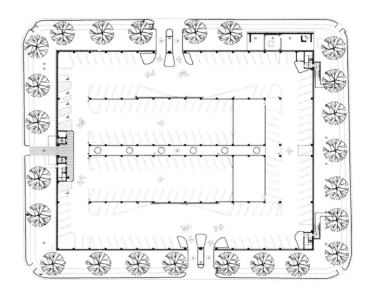

1 电梯	1 Elevator
2 中庭	2 Atrium
3 楼梯	3 Stairs
4 汽车入口	4 Automobile Entry
5 汽车出口	5 Automobile Exit

总平面图 SITE PLAN

2号停车场，Elliott + Associates Architects，美国，俄克拉何马州，俄克拉何马城，2011年

建筑表皮是一层不锈钢网格，开放的编织方式使光和空气都能进入建筑内，同时从室外看时，能够反射俄克拉何马的阳光。停车场内部，建筑师在楼梯间使用了同样的彩色荧光灯来强调色彩和楼层的衔接，为使用者营造了有趣的氛围。所有的内墙均为白色，净高度0.15~3.05 m 不等，以醒目的图案、彩色的荧光灯营造出彩色的楼层，容易区分。

Car Park Two, Elliott + Associates Architects, Oklahoma City, Oklahoma, USA, 2011

The building is wrapped in a layer of stainless metal mesh. The open weave allows light and air to enter the building and reflects the light in Oklahoma when seen from outside. The architects added the same colored fluorescent lighting in the stairwells to emphasize the color / level connection and a fun atmospheric journey. All interior surfaces are painted white and the clear height is 10' – 6". Color-coded levels are defined with bold graphics and colored fluorescent lighting visually reinforces the level you decide on.

© Scott McDonald

材料：不锈钢、彩灯

Materials: stainless steel, colored lighting

© Scott McDonald

行云流水
SMOOTH LIKE FLOATING CLOUDS AND FLOWING WATER

 黄土赭色 Sienna

月白 Moon White

I' Park 城样板房，UNStudio，韩国，水原，2008年

I' Park 城样板房是一个展厅，展示了住宅楼的城市设计和外观设计。白色外墙与深色玻璃相得益彰，使流畅的线条更加突出，跃然纸上。该设计的主要意图是通过精心设计的色彩，带给参观者全新的体验。从大楼入口开始，参观者沿游览路线，一边行进一边参观，这是一个"行进中的展览"。外观拥有完美流畅的线条，仿佛是雕塑家塑造出来的优雅雕塑，此起彼伏。

Model House for I' Park City, UNStudio, Suwon, South Korea, 2008

The principle design intention was to curate the visitor experience by elaborating on an implicit circulation strategy. The complement of white wall and dark glass further highlights the fine lines and makes the building full of life. The route of the visitor, from the approach to the building and throughout the tour within, is treated as an ongoing exhibition. The perfect smooth lines of the façade rise and fall like the graceful works of the sculptors.

材料：混凝土、玻璃

Materials: concrete, glass

© Christian Richters

© Christian Richters

© Iwan Baan

蜜色 Honeydew

亮天蓝色 Light Sky Blue

中紫色 Medium Purple

弱紫罗兰红 Pale Violet Red

I' Park 城样板房，UNStudio，韩国，水原，2008年

　　室内运用淡淡的紫色局部粉饰，如同梦境一般，仿佛我们都在追寻着我们自己编织的白色和紫色的梦，来回穿梭在不同想象的梦境中。一条相同的小路平行地将室内展示内容与相邻的大片地区和景观相连，这里是未来发展的区域。游览路线强调了"回家的体验"的重要性，这是整个I' Park城城市发展的组织驱动力，是整体品牌战略的核心。

Model House for I' Park City, UNStudio, Suwon, South Korea, 2008

The interior is partly embellished with light purple like a fairyland. It's as if we are pursing our dreams woven with white and purple, and shuttling between different dreamlands. In parallel, the same path oscillates between a focus on the interior exhibited content and the adjacent larger site and landscape, which is the location of the future development. The circulation route underscores the importance of the "coming home experience" which is the organizational driver of the entire I-Park City Urban development, and the core of the overall site branding strategy.

材料：混凝土、玻璃

Materials: concrete, glass

© Iwan Baan

© Iwan Baan

高雅购物天堂
ELEGANT SHOPPING PARADISE

 熏衣草淡紫 Lavender

 亮钢蓝 Light Steel Blue

 银灰色 Silver

 蓝紫罗兰 Blue Violet

 淡绿色 Light Green

格洛斯楚普购物中心,C. F. Møller Architects,丹麦,哥本哈根,格洛斯楚普,2010年

紫色,一直是优雅的代名词。用淡紫色调构建的购物中心全新的立面更给人高雅之感。在蓝色天空的影响下,立面又映射出淡淡的蓝色,平添了几分沉静大气之感。当夜幕降临,LED灯衬照着立面,整座购物中心在黑暗中熠熠闪耀。动感彩色的灯光被编成不同的序列,为建筑营造出变幻多姿的外观。淡紫色加上曲线的设计,让原本平淡的建筑焕然一新,使之成为高雅的购物天堂。

Glostrup Storcenter, C. F. Møller Architects, Glostrup, Copenhagen, Denmark, 2010

Purple has always been the synonym for grace. Composed of purple sheets, the new façade of Glostrup Storcenter presents a graceful appearance. Influenced by the blue sky, the façades reflects a shade of blue, and a calm and grand atmosphere is created. In the evenings the façade is back-lit by LED lamps, by which the shopping center makes a distinctive impression in the darkness. The colored lighting is dynamically programmed in different sequences to create an ever-changing appearance. With the involvement of lavender and the delicate curves, the pain building took on a completely new look and therefore became an elegant shopping paradise.

材料:聚碳酸酯板

Material: polycarbonate sheets

© Steen & Strøm

© Steen & Strøm

© Steen & Strøm

© Steen & Strøm

紫之梦幻 | PURPLE

彩翼
THE COLORFUL WINGS

 金色 Gold

 粉蓝色 Powder Blue

 弱紫罗兰红 Pale Violet Red

 钢蓝 Steel Blue

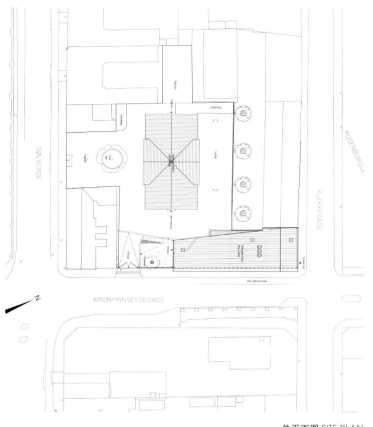

总平面图 SITE PLAN

© Adam Mørk

索夫盖德学校，C. F. Møller Architects，丹麦，哥本哈根，2012年

学校的设计与周边建筑的规模和走向十分吻合，并从山墙装饰中汲取灵感。靠近Kronprinsessegade街的立面效仿了Dronningegården住宅楼的风格，在较高楼层采用了内嵌式设计，制造出一种幽深而宽阔的双层立面的错觉。同样的设计手法被应用在了朝向校园的立面上。该立面在内凹处涂以缤纷的色彩，为建筑带来了勃勃生机。同时，这种设计还突显了立面纵深的空间感。这些色彩涂饰在建筑表皮的抹灰上，与Kronprinsessegade街道旁的古典主义建筑的彩色灰泥立面相得益彰。

Sølvgade School, C. F. Møller Architects, Copenhagen, Denmark, 2012

The design of Sølvgade School is adapted to the sizes and directions of the surrounding buildings, and draws inspiration from their gable motifs. Towards Kronprinsessegade, one façade of the school is modelled on Dronningegården, which has recessed façades on its upper floors, thereby creating the illusion of a deep, spacious double façade. In the school façade, the practice has added extra life to a similar double façade by providing the internal rebates with a wide palette of colors. This also further emphasizes the deep spaciousness of the façade. The colors are set in the building's exterior rendering, and refer to the painted or stained plaster façades of the Classicist buildings along Kronprinsessegade.

材料：玻璃、涂料、石膏彩绘

Materials: glass, paint, plaster painting

© Adam Mørk

© Adam Mørk

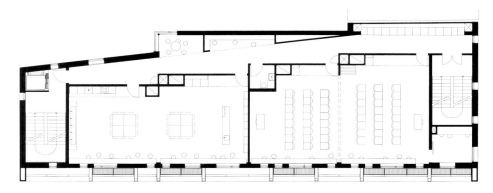
平面图 PLAN

    靛青 Indigo

    皇家蓝 Royal Blue

    暗蓝色 Dark Blue

    浅海洋绿 Light Sea Green

    金色 Gold

    金菊黄 Goldenrod

© Adam Mørk

索夫盖德学校，C. F. Møller Architects，丹麦，哥本哈根，2012年

建筑内部的配色方案有助于为学生提供生机勃发、色彩缤纷的空间体验，但并无压抑感。在室内，只有立面后的背光墙面涂有色彩。为了避免向阳墙面反射光线所带来的颜色，因此垂直于日光入射面的墙面上没有粉刷色彩。朝向街道和庭院（建筑表皮）的墙壁则粉刷成了白色，因此教室里只有一面"活跃"的彩色墙。这面墙还容纳了技术设施、固定装置、门、衣帽储存空间等，而其他空间则采用了中性的色彩。这样的组合使得建筑的配色方案兼具动感和沉静，为学生创营造出了动感十足的学习环境。

Sølvgade School, C.F. Møller Architects, Copenhagen, Denmark, 2012

The interior color scheme helps to provide a vibrant and coloristic experience of the spaces, without totally dominating them. Inside, only the walls back-lit by the façade have been colored. Walls lying perpendicular to the incidence of daylight have not been colored, to avoid the space taking on the color of the reflected light from the sunlit side walls. The walls towards the street and courtyard (the building envelope) are white, with the result that there is only a single "restless" colored wall in the classrooms. This single wall also holds the technical installations, fixtures, doors, cloakroom spaces, etc., so that the rest of the space is neutral in color. In combination, this creates a color scheme that is simultaneously vibrant and calm, and provides a dynamic learning environment.

材料：橡胶、涂料

Materials: rubber, paint

© Adam Mørk

弦外之音
OVERSTONE

 弱绿宝石 Pale Turquoise

 暗黄褐色 Dark Khaki

 绿黄色 Green Yellow

 银灰色 Silver

 中紫色 Medium Purple

 深红色 Dark Red

 深洋红 Dark Magenta

MUMUTH音乐剧院，UNStudio，奥地利，格拉茨，1998－2008年

　　MUMUTH音乐剧院归属于奥地利格拉兹音乐和表演艺术大学，为年轻音乐家提供音乐和表演教育基地。因此，建筑师尽可能使建筑富有音乐性。外立面银色的网是由一个一个的音乐梦想编织而成的，每一个音乐梦想都是一个小宇宙等着强力爆发。螺旋结构的建筑元素使大厅成为自由流畅的空间，连接着入口和礼堂以及楼上的音乐室。此外，建筑师还发现了之前一直忽略的另一元素——重复。建筑师在设计中融入重复的图案，通过多种途径应用在建筑立面上，以达到预期效果。金属质感的楼梯、深红色的铺面、优雅的装饰墙面、淡紫色的灯光缓缓萦绕在周围……每个细节的呈现，都让你沉醉不已，深陷其中，感受色彩与音乐带给心灵的慰藉与释放。

MUMUTH Music Theater, UNStudio, Graz, Austria, 1998-2008

The MUMUTH theatre belongs to the University of Music and Performing Arts Graz and is a place where young musicians receive their instruction in the performing and musical arts. Therefore the desire to make a building that is as much about music as a building can be. The glittering mesh façade seemed to be woven with countless music dreams, each of which is like a small universe waiting for a powerful outbreak. The free-flowing space of the foyer is made possible by a spiraling constructive element that connects the entrance to the auditorium and to the music rooms above. In addition, the architects learned that there is another element that they had not seriously studied before: the element of repetition. They used a repetitive pattern, of their own design, and applied this to the façades in various ways to achieve the desired effects. The metal stairs, scarlet facing, elegant walls and lavender lighting spread around, intoxicating you, and you will feel the comfort and relief brought by colors and music.

© Christian Richters

© Christian Richters

材料： 室外 钢材、金属网、玻璃
　　　　室内 混凝土、木材、钢材

Materials: Exterior: steel, metal mesh, glass
　　　　　　Interior: concrete, wood, steel

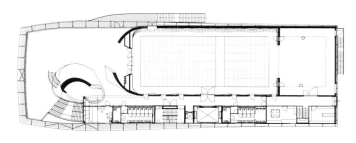

一层平面图 FIRST FLOOR PLAN

© Christian Richters

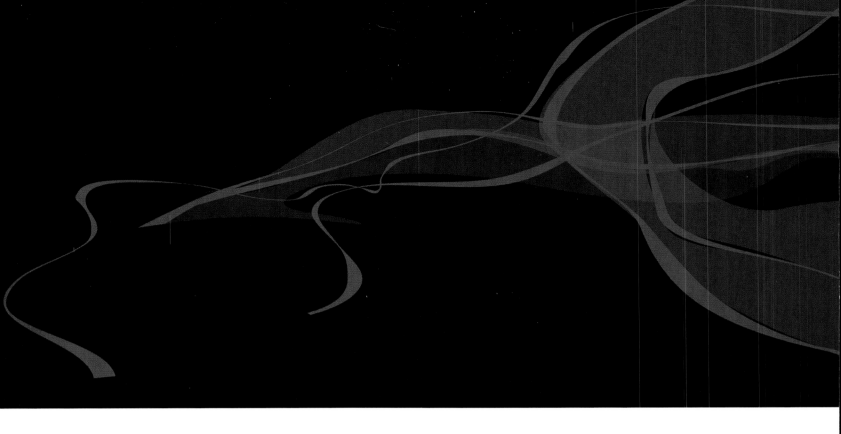

■西方有石名黛，可代画眉之墨。——《红楼梦》■黑色，象征神秘、未知、深遂、力量、权威、肃穆，但是也代表消极和死亡。■黑色和白色一样，作为一种万能色常出现在建筑局部，但以黑色为主色或者全色的建筑却较其他色少见。一旦使用，对设计师把握光线和质感的能力有着极高的考验。■黑色的建筑强调体块，吸收并保存能量，有着神秘的力量。■其实在建筑中，纯黑是很少单独使用的。在"光强色强，光弱色弱"的规律下，很多空间里的黑色其实是因为光少而产生的。因此我们将一些灰色和较深颜色的运用，也放入这章中一并解读。

■ There is a saying in Chinese classical novel Dream of the Red Chamber that "somewhere in the West there is a mineral called 'dai' (a name for black in Chinese) which can be used instead of eye-black for painting the eyebrows with." ■ Black stands for mystery, unknown, abstruse, power, authority and solemnness; while on the other side, it also represents negativity and death. ■ Like white, black is also a almighty color, which appears in most architecture. However, black is seldom used as the main color or the only color for the architecture. It will pose a challenge to the architects to manage light and texture once black is used. ■ Black architecture emphasize the volumes, and they can absorb and store energy, having a mysterious power. ■ Actually, pure black is rarely used alone. According to the rule of "intense light brings strong colors while weak light brings faint colors", black is seen in many spaces due to a lack of light. We therefore include gray and some other darker colors used in architecture in this chapter.

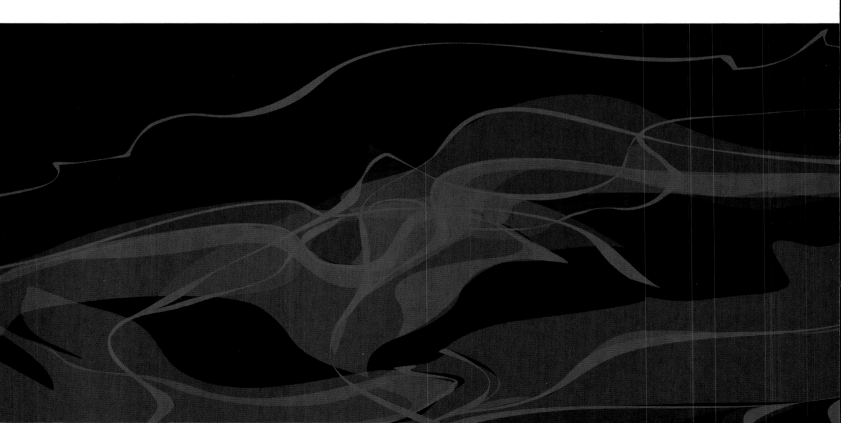

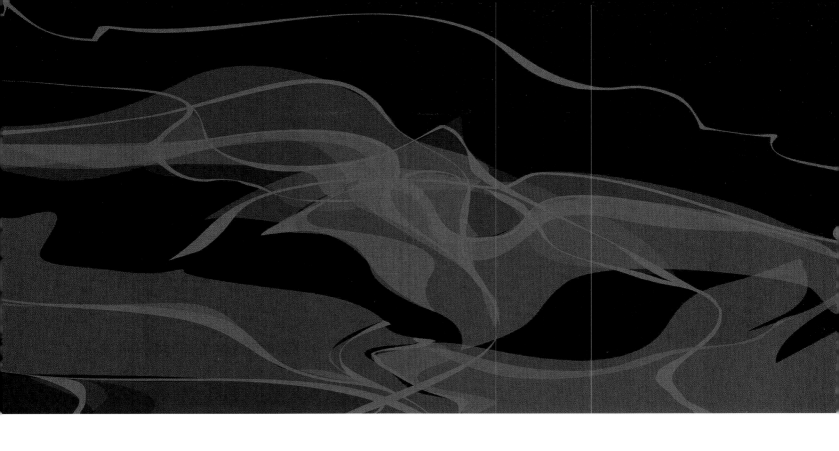

黛之神秘
BLACK
207

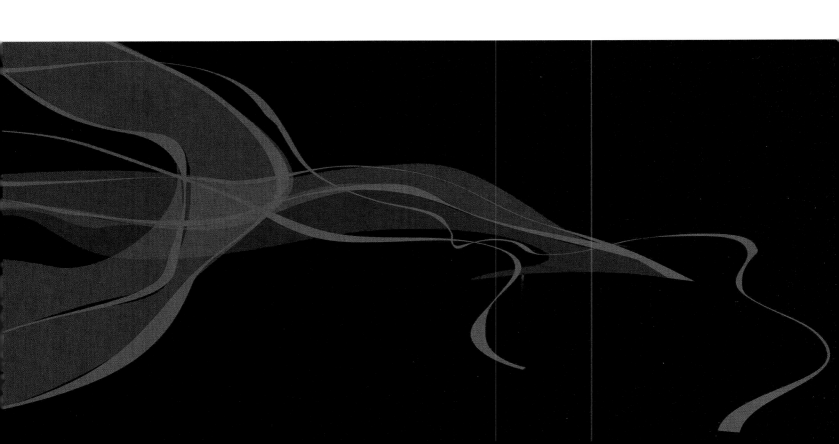

黑白时尚
BLACK AND WHITE FASHION

雪白色 Snow

 浅灰色 Light Gray

 暗淡灰 Dim Gray

黄土赭色 Sienna

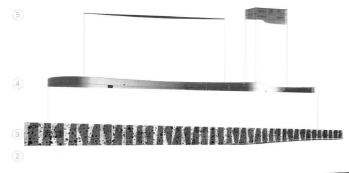

展开图 FOLDOUT DIAGRAM

© Pedro Pegenaute

© Pedro Pegenaute

索里亚商场，Vaîllo & Irigaray，西班牙，纳瓦拉，塔乔纳尔，2006年

  Soria.com 商场所在的位置地势绝佳，其设计源于乡村农业建筑。它像石凳镶嵌在地面上。建筑的外墙由U形玻璃构筑，柔美、轻盈、起伏有致。它沿着建筑石墙上的飞檐蜿蜒伸展，将一个个分散的区域在顶端围拢。

  护墙板就像一块巨大的帆布，呈褶皱状延展，上面的孔隙（空气、海绵）构成了概念性图案，仿佛一个巨大的都市货架，商店的后部变成了建筑的外立面）。

  顶部由半透明玻璃构成的外墙蜿蜒盘绕，将内部各个区域包裹成了一个个巨大的气泡。

Soria.com, Vaîllo & Irigaray, Tajonar, Navarra, Spain, 2006

 Soria.com is a building located in a strong topography which takes its origin from the rural agricultural architecture: Stone benches, molding the ground. A soft, light and undulating enclosure, made by means of U-glass elements, goes along the upper cornice of the building on the stone wall and thus, dismounts the wall to organize the scattered areas of the building in the vertex opposite the wall.

 Baseboard: a huge canvas pleated fabric, conceptually plotted with holes (air, sponge); it is equivalent to a huge urban shelf (the back parts of the shops become façades).

 Crowning: a winding enclosure of translucent glass (U-glass) packs in a quantum way the packaged programs into big urban bubbles.

材料：织物、玻璃

Materials: fabric, glass

外表朴实、内有乾坤
A SLIENT BOX WITH WONDERFUL CONTENTS HIDDEN INSIDE

绿黄色 Green Yellow

浅灰色 Light Gray

暗淡灰 Dim Gray

纯黑 Black

总平面图 SITE PLAN

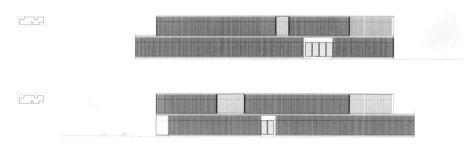

立面图 ELEVATIONS

流浪者之家，Javier Larraz，西班牙，纳瓦拉，潘普洛纳，2010年

黑色是一种强大的色彩，具有神秘、黑暗、暗藏力量的含义。它庄重而高雅，可以让其他颜色突显出来。因此被黄色灯光渲染的入口格外醒目，吸引你前去一探究竟。建筑好似一个沉寂的盒子，将万千精彩内藏其中，引起观者的好奇心。同时，它恰当地缩小了自身的比例，融于荒芜的半城市化环境之中。由铝材构成的外部格栅保证了使用者的隐私性，同时避免了可能发生的安全问题。建筑呈现了匀称统一的外观，其规模与环境相得益彰。

Shelter Home for the Homeless, Javier Larraz, Pamplona, Navarra, Spain, 2010

Black is a powerful color which contains mysterious, dark hidden strength. It also has a solemn and graceful side and can emphasize other colors. Illuminated by the yellow lighting, the striking entrance attracts visitors to go inside. A silent box is proposed, that protects its contents from the curiosity of the onlookers, and that adequately integrates its reduced scale in a semi-urban, bleak environment. An exterior lattice conformed by aluminum profiles guarantees the desirable privacy of the users, and at the same time resolves the possible intrusion problems that may occur in such a center, and configures an homogeneous and unitary exterior image, adapting the scale of the building to its environment.

© Iñaki Bergera

材料：铝、混凝土、U型玻璃

Material: aluminum, concrete, U-glass

© Iñaki Bergera

神秘波浪
MYSTERIOUS WAVES

 暗黄褐色 Dark Khaki

 亮菊黄 Light Goldenrod Yellow

 深橙色 Dark Orange

 中海洋绿 Medium Sea Green

 深洋红 Dark Magenta

 闪兰色 Dodger Blue

 靛青 Indigo

 纯黑 Black

"空间演化"展览，UNStudio，德国，法兰克福，2006年

　"空间演化"展览中展出了UNStudio在过去18年来的建筑设计作品。展览的空间通过黑色的展台和彩色的波浪线，充分调动人的视觉神经。这些彩色的波浪线也可以将空间分割，组织布局。此次展览并非按照展品的类型或时间来陈列，而是按照自身发展路线陈列。整个展厅被设计成一个三维图形，展示出UNStudio创造的5种不同的设计标准相关的工作进展状况。

Exhibition Evolution of Space, UNStudio, Frankfurt, Germany, 2006

The exhibition "Evolution of Space" shows the products of 18 years of architectural practice, arranged not according to typology or chronology, but according to a more personal line of development. The black exhibition shelters colored wavy lines fully excite people's visual nerve. The gallery space has been converted into a three-dimensional graphic, charting the progress of the work in relation to five different design standards invented by UNStudio. Furthermore, the wavy lines can also serve to divide spaces and arrange the layout.

材料：木材、织物

Materials: wood, fabric

© Uwe Dettmar

© Uwe Dettmar

© Uwe Dettmar

神秘黑色住宅
HOUSING IN MYSTERIOUS BLACK

雪白色 Snow

暗瓦灰色 Dark Slate Gray

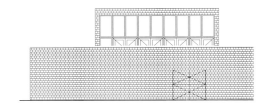

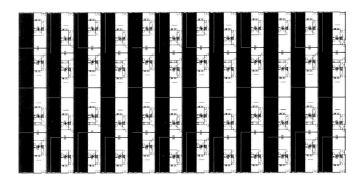

一层平面图 GROUND FLOOR PLAN

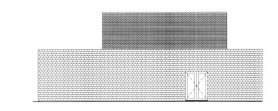

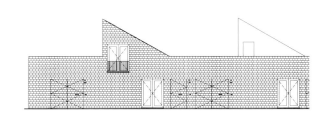

立面图 ELEVATIONS

© Rob't Hart

© Rob't Hart

天井岛,MVRDV,荷兰,海牙,2005年

这一片新住宅区包含了48个天井式住宅单元（高4.5 m），且所有住宅都被安置在一座岛屿上。房子的外墙和屋顶覆盖着黑色的木瓦，突出了内弯、雕塑形态的重量感。房门也覆盖着木瓦，与墙面在视觉上保持一致，并能够保证隐私。另外，采用折叠门能够保证隐私，同时还能保证向外的视野。

Patio Island, MVRDV, Den Haag, Holland, 2005

This new residential area, comprising 48 patio style dwelling units (h 4.5 m), is confined to just one island. The façades and roofs are clad with black shingles, accentuating the heaviness of the introverted, sculptural form. The doors are also clad in shingles so as to leave the wall visually intact and to ensure the privacy of the house. Executing the doors in stable-door style combines privacy with an outward view.

材料：木瓦

Material: shingles

行走的雕塑
MOVING SCULPTURE

雪白色 Snow

亮菊黄 Light Goldenrod Yellow

弱绿宝石 Pale Turquoise

黄土赭色 Sienna

灰色 Gray

暗淡灰 Dim Gray

Masrah Al Qasba剧院,Magma Architecture,阿联酋,沙迦,2012年

进入Masrah Qasba剧院300席的会堂里,观众会发现自己被雕塑般黑色波浪起伏的空间所包围,让人联想到沙迦的风景,异常的壮观。照明设施都隐藏在可延展的织物表层之下,提高了空间的连续感。这座剧院的设计根植于沙伽的自然和历史要素,同时契合文化的未来发展。在游牧时代,沙漠上的人们都是通过口述传递日常生活的消息。剧院就像一个沙漠景观的透视图,折射出沙伽的景色。弯曲的表层和分散在顶棚上的灯光条带就像是傍晚的阳光照射到了沙丘上,进入会堂的观众犹如置身红尘之外,进入一个迷幻的舞台空间。

Masrah Al Qasba Theater, Magma Architecture, Sharjah, UAE, 2012

Visitors to the 300 seat auditorium find themselves wrapped in a sculpturally undulating space evoking recollections of the landscapes of Sharjah. Carefully concealed lighting elements shining through stretchable fabric surfaces enhance the spatial sensation of continuous enclosure. The theater was rooted in the nature and history of the Emirate of Sharjah as embracing its cultural future. In the days of nomadic life news was transmitted through oral narrative in the open-air of the nocturnal desert landscape. The design of the Al Qasba auditorium refers to the nature of Sharjah as inspiration and suggests an inverted, fully enclosing scenographic landscape. Undulating surfaces with light strips on fold lines scatter across the ceiling evoking images of evening sun streaking sand dunes. Visitors entering the auditorium disconnect from everyday life to immerse into the surprising and striking stage.

材料:聚酰胺和氨纶纺织物、铝合金型材

Materials: polyamide and spandex textile, aluminum profile

© Torsten Seidel

© Magma

扭曲的黑盒子
TWISTED BLACK BOX

雪白色 Snow

 灰菊黄 Pale Goldenrod

 暗淡灰 Dim Gray

 黄绿色 Yellow Green

中绿宝石 Medium Turquoise

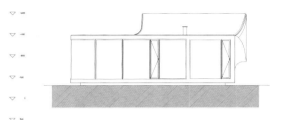

南立面图 SOUTH ELEVATION

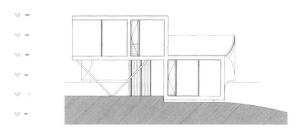

北立面图 NORTH ELEVATION

© Christian Richters

NM别墅，UNStudio，美国，纽约州北部，2007年

NM别墅的概念是基于一个分叉的黑色盒子，当把盒子的平面和剖面扭转，这个鞋盒状建筑便分成两个错层的独立体量，建筑的侧墙上装有落地窗，阳光可透过窗子进入室内。从外面看，房子的颜色像大地，窗子则像黄昏的天空，建筑表面反射出周围的风景。

VilLA NM, UNStudio, Upstate New York, USA, 2007

The conceptual model for VilLA NM is a box with a blob-like moment in the middle; a twist in both plan and section that causes a simple shoebox to bifurcate into two separate, split-level volumes. The side walls consist of floor to ceiling windows that let in the light. On the outside, the house is colored like the earth; its windows are tinted like the sky at dusk. The building captivates the surrounding landscape.

材料：水泥、玻璃、木材、金属

Materials: cement, glass, wood, metal

© Christian Richters

浅灰色 Light Gray

 鹿皮色 Moccasin

 暗淡灰 Dim Gray

NM别墅，UNStudio，美国，纽约州北部，2000–2007年

　　NM别墅内部通体洁白、平滑，属于未来派建筑，是久居都市、追求理想生活的人们渴望的一处"田园雅居"。周围风景不断发生着变化，身处其中你会感觉自己就是这风景的一部分。

Villa NM, UNStudio, Upstate New York, USA, 2000-2007

All white and smooth, Villa NM is a futuristic hut. It is a rural retreat for a family of idealistic, glamorous urbanites. In the house, you will feel you are part of the landscape that changes around you perpetually.

材料：胶合板、石膏、聚碳酸酯、环氧树脂、玻璃、木材

Materials: plywood, plaster, polycarbonate, epoxy, glass, wood

© Christian Richters

© Christian Richters

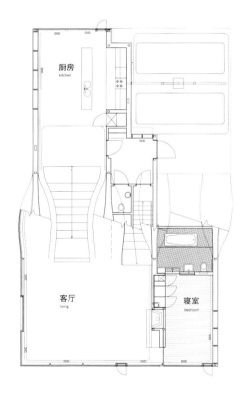
一层平面图 FIRST FLOOR PLAN

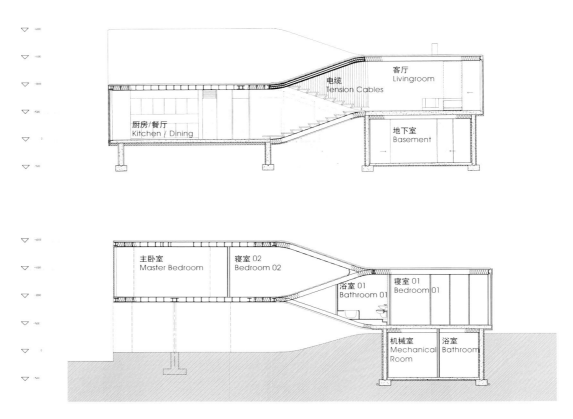

剖面图 SECTIONS

© Christian Richters

黛之神秘 | BLACK | 221

折叠的壳
FOLDED SHELL

 金菊黄 Goldenrod

 暗淡灰 Dim Gray

 橙红色 Orange Red

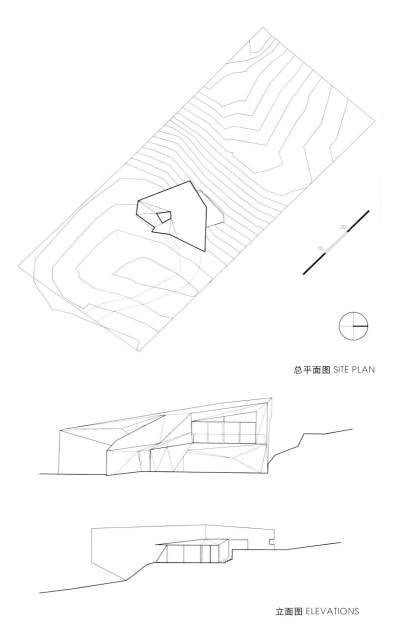

总平面图 SITE PLAN

立面图 ELEVATIONS

© John Gollings

© John Gollings

克莱因瓶住宅，McBride Charles Ryan，澳大利亚，莫宁顿半岛，2008年

以纯几何的眼光拆开来看，建筑是由若干三角面和不规则四边形拼接而成的一个能够居住的封闭空间。整体结构非常复杂，没有任何一组相似的几何结构出现。建筑的基本用色也很简洁，经典的黑白色搭配在不经意间散发出优雅气息。

Klein Bottle House, McBride Charles Ryan, Mornington Peninsula, Australia, 2008

Analysed in a pure geometrical way, the building is an enclosed living space that consists of several triangle planes and trapezoids. The structure is so complex that there are no similar geometries in the house. The basic color of the building is simple, too. The classic black and white easily gives a sense of elegance.

材料：压缩纤维水泥板

Materials: compressed fiber cement sheet

© John Gollings

克莱因瓶住宅，McBride Charles Ryan，澳大利亚，莫宁顿半岛，2008年

从平面上看，住宅的内部空间比较简单。进入住宅，便是红色的折转楼梯，一层的主要房间都围绕楼梯外侧旋转布局，楼梯中心围绕成中心庭院。窗户的设计同样遵守基本几何元素，为环绕的内部空间提供了采光和敞开的空间感。楼梯部分可以理解为克莱因瓶探出的瓶颈，贯穿了整个瓶体结构。室内空间除了在交流区域的楼梯采用的活跃的大红色作为主色外，其他空间仍主要以恬适的暖色调为基色，保证温馨舒适的居住感受，家具则仍然保持黑白的对比设计。

Klein Bottle House, McBride Charles Ryan, Mornington Peninsula, Australia, 2008

In plane view the inner space of the house is rather simple. The red staircase firstly jump into sight when people enter the house. Most of the rooms on the ground level circle around the staircase which forms a courtyard in the center. Windows are also designed as geometrical elements and they provide illumination and open space to the interior. The staircase can be interpreted as the neck of the Klein bottle, which runs through the entire structure. Despite of the staircase in the communication area with red as its main color, other spaces are tinted mainly with sweet, warm colors to guarantee home comfort. The contrast between black and white continues to appear on furniture.

材料：数控铣胶合板、石膏板

Materials: CNC-milled plywood, plaster board

© John Gollings

雪白色 Snow

 硬木色 Burly Wood

 纯红 Red

 暗淡灰 Dim Gray

© John Gollings

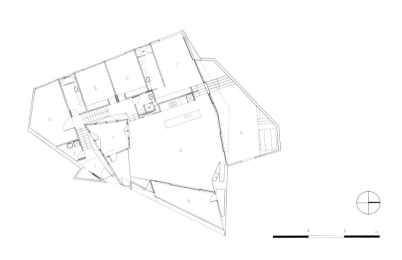

1 门厅	1 Foyer
2 庭院	2 Courtyard
3 洗衣房	3 Laundry
4 娱乐室	4 Rumpus
5 寝室 1	5 Bedroom 1
6 寝室 2	6 Bedroom 2
7 主卧室	7 Master Bedroom
8 客厅/餐厅	8 Living / Dining
9 东侧露天平台	9 East Deck
10 北侧露天平台	10 North Deck

平面图 PLAN

© John Gollings

黛之神秘 | BLACK | 225

迷幻银盒
MAGIC SILVER BOX

 银灰色 Silver

 水鸭色 Teal

 军兰色 Cadet Blue

 暗紫罗兰 Dark Violet

 纯红 Red

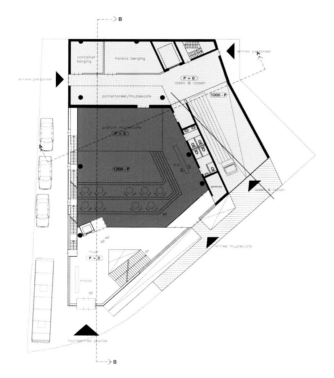

平面图 PLAN

© Jeroen Musch

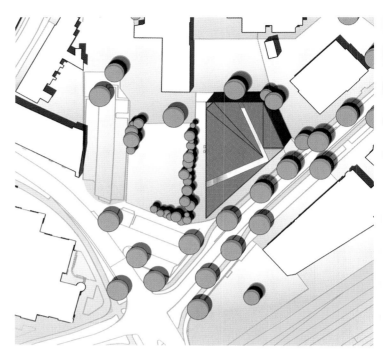

总平面图 SITE PLAN

流行音乐厅，de Architekten Cie.，荷兰，希尔弗瑟姆，2010年

这座新建筑的建造结构显示出其作为音乐厅的两个主要功能厅：演奏厅和音乐咖啡厅。根据其功能需要，演奏厅为封闭式空间，而音乐咖啡厅则呈现出较为开放的氛围。建筑立面采用几何建筑美学原理，外表皮使用银色来呈现音乐厅的律动。时尚的新音乐厅在空间和建筑体系上的表现手法展现了"盒中盒"的设计理念。同时，这样的规划提升了君王般的崇圣感。不同颜色的灯效在划分功能区域的同时，把空间渲染得绚丽而迷幻。

The Sovereign Pop Venue, de Architekten Cie., Hilversum, the Netherlands, 2010

The construction program of The Sovereign shows that the new building is typified by two main functions: the main auditorium and the music café. In line with its function, the main auditorium has a closed character, whereas the music café has a more open ambience. The façade was based on the principle of geometrical aesthetic by using a silver skin to present the rhythm of the Pop Venue. This layout simultaneously generates The Sovereign's image: the spatial and architectonic representation of the new pop venue is fashioned in the principle of a box within a box. The lighting in different colors divides the functional zones and renders the interior as a fascinating space.

材料：金属板、玻璃

Materials: metal panels, glass

为了忘却和理解的纪念
A COMMEMORATION FOR FORGET AND UNDERSTANDING

 钢蓝 Steel Blue

 亮蓝灰 Light Slate Gray

 亮天蓝色 Light Sky Blue

 深灰色 Dark Gray

 纯黑 Black

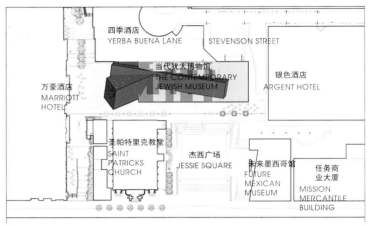

总平面图 SITE PLAN

© bitterbredt

当代犹太博物馆,Daniel Libeskind Studio,美国,旧金山,2008年

从外观上看,该建筑坚固、紧实,却又充满了隐喻和矛盾的色彩。建筑外壳由深灰色钢板包裹,折线形体来源于希伯来文字母"chet"和"yut",这两个单字构成了词语"L'Chaim",意思为"生命"。建筑像利刃一般插入有着新古典风格的红砖墙立面的矩形框架结构空间中(即原来的旧金山地区的发电站),使这两个看起来毫不相干的形体不可思议地结合在一起。事实上设计师依然在用非常现代的结构形体表现非常传统的、原始的犹太文化和精神。

Contemporary Jewish Museum, Daniel Libeskind Studio, San Francisco, USA, 2008

The building presents a solid, even tense appearance. Yet it is full of metaphors and contradicts. The volume of the building is wrapped in dark gray steel panels. It is based on the Hebrew letters "chet" and "yut", forming the word "L'Chaim", which means "To Life". It like a sharp knife thrusting into the rectangle neoclassic framed structure, which has a façade composed of red bricks (the former power station in San Francisco). The two seemingly irrelevant blocks were incredibly combined. In fact, the designers are expressing traditional and original Jewish culture and spirit with modern structures.

材料:钢板、玻璃

Materials: steel panels, glass

© bitterbredt

银灰色 Silver

 柠檬绸 Lemon Chiffon

 暗淡灰 Dim Gray

 纯黑 Black

© bitterbredt

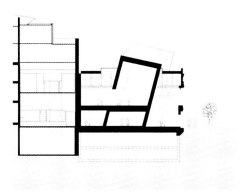
横向剖面图 CROSS SECTION

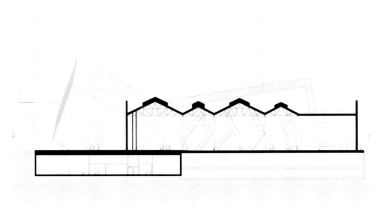
纵向剖面图 LONGITUDINAL SECTION

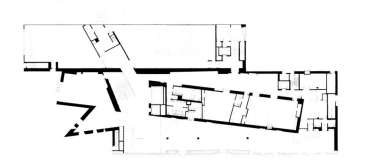
一层平面图 GROUND FLOOR PLAN

当代犹太博物馆，Daniel Libeskind Studio，美国，旧金山，2008年

博物馆的内部依然延续着强烈的暗示和隐喻。从主入口进入门厅，人们瞬间被拉入了一个奇妙的空间中，人工照明与自然光交织出模糊、感性而又带有一丝神秘感的氛围。整个建筑中没有一面垂直的墙体或一个直角的空间，建筑形体中隐藏的"生命"和"天堂"的寓意是对犹太文化的生命力和未来的肯定与赞美。或许，选择忘却与理解，不是回避，不是逃离，而是对更加自信地生活的纪念。

Contemporary Jewish Museum, Daniel Libeskind Studio, San Francisco, USA, 2008

Hints and metaphors continue to appear in the interior of the museum. Stepping into the hallway through the entrance, visitors are immediately dragged into a magic space. Artificial illumination together with natural light creates a obscure, emotional and mysterious ambience. There is no single perpendicular wall or orthogonal space in the building. "Heaven" and "life" are hidden in the form of the building, implying the vitality of Jewish culture and affirmation and praise for the future. Perhaps choosing to forget and understand instead of to avoid or escape is the commemoration of a more confident life.

材料：钢板、玻璃

Materials: steel panels, glass

坚实的守护
STAUNCH PROTECTION

 浅灰色 Light Gray

 纯黄 Yellow

 橙色 Orange

 纯红 Red

 纯绿 Green

 纯蓝 Blue

 深灰色 Dark Gray

 灰色 Gray

总平面图 SITE PLAN

© Invisible Gentleman

© Invisible Gentleman

Braamcamp Freire中学，CVDB Arquitectos，葡萄牙，里斯本，庞丁哈，2012年

学校建筑的立面特色鲜明，由原地浇筑的清水混凝土墙壁和混凝土板构成，该设计有助于减少维护费用。混凝土面板经过精心设计与立面的朝阳性相得益彰。混凝土的冷灰色调使立面看起来沉静而庄严。靠近窗户的混凝土表皮粉刷了各种不同的颜色，使立面呈现反射，充满生机勃勃之感。

Braamcamp Freire Secondary School, CVDB Arquitectos, Pontinha, Lisboa, Portugal, 2012

The façades of the school are characterized by fair faced concrete walls built in situ and precast concrete panels. This solution contributes to decreasing maintenance costs. The concrete panels were carefully designed to respond adequately to the façades' solar orientation. The gray tonality of concrete makes the façade look calm and grand. The concrete surfaces close to the windows were painted in different colors, allowing the occurrence of vibrant reflections in the façades.

材料：混凝土

Material: concrete

© Invisible Gentleman

黛之神秘 | BLACK

纯黄 Yellow

纯红 Red

纯蓝 Blue

灰色 Gray

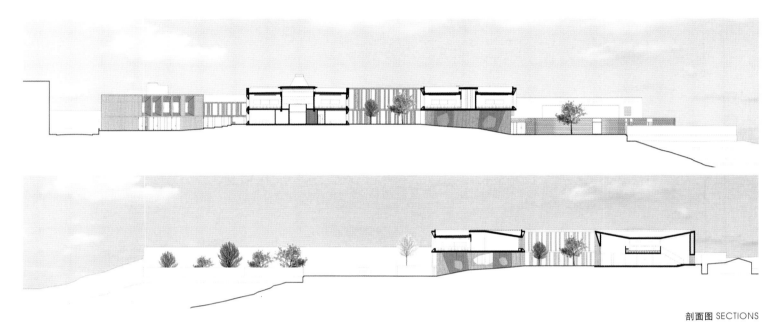

剖面图 SECTIONS

© Invisible Gentleman

© Invisible Gentleman

© Invisible Gentleman

© Invisible Gentleman

© Invisible Gentleman

Braamcamp Freire中学，CVDB Arquitectos，葡萄牙，里斯本，庞丁哈，2012年

内部空间选用了耐磨材料以达到集约利用和降低维护成本的目的。在流通空间，墙壁由混凝土块构成，上方包含吸声混凝土块。沿着流通空间和社交空间，可看到色彩鲜艳的壁龛。多功能厅由木质龙骨和吸声板装饰而成。不规则形状的门洞、竖条状光带、盘旋而上的楼梯……到处都蕴含着设计的灵感。

Braamcamp Freire Secondary School, CVDB Arquitectos, Pontinha, Lisboa, Portugal, 2012

In the interior spaces, adequate resistant materials were chosen for an intensive use and low maintenance costs. In the circulation spaces, the walls are built with concrete blocks and on the upper part include concrete acoustic blocks. Along the circulation and social areas can be found niches painted in strong colors. The multi-purpose hall has timber studs and acoustic panels. Irregular shaped door openings, vertical light bands, spiral stairs…inspiration is found anywhere.

材料：混凝土、木材、涂料

Materials: concrete, wood, paint

© Invisible Gentleman

黑暗中的水晶城堡
CRYSTAL CASTLE IN DARKNESS

 亮天蓝色 Light Sky Blue

 绿黄色 Green Yellow

 浅灰色 LightGray

雪白色 Snow

松恩&菲尤拉讷艺术博物馆,C.F. Møller Architects,挪威,弗德,2012年

新博物馆的建筑设计融入了周围的独特景观。博物馆仿佛如晶亮的冰块自周边高山上滑落下来。水晶般透明的造型呈现不对称平面布局,而立面则形成位移变化。建筑立面覆有白色玻璃,装饰有倾斜的线条网格,让人联想到冰块的断裂线。另外,线条网格标示出不规整的窗缝。夜晚,这些线条被点亮,使博物馆看起来就像是在黑暗中闪闪发光的水晶。

Sogn & Fjordane Art Museum, C.F. Møller Architects, Førde, Norway, 2012

The new museum draws upon the distinctive landscape for its architectural expression: the museum lies like a crystal-clear block of ice that has slid down from the surrounding mountains. The crystalline form provides an asymmetrical plan solution, with varying displacements in the façade. The façade is clad in white glass with a network of angled lines, reminiscent of the fracture lines in ice. This network also defines the irregular window apertures. In the evening these lines are illuminated, so that the museum lies like a sparkling block in the middle of the town's darkness.

材料:玻璃

Material: glass

© Oddleiv Apneseth

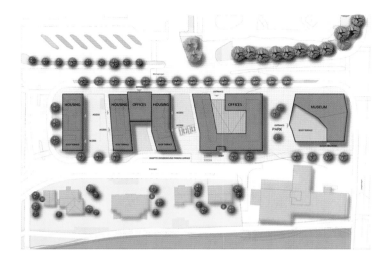
总平面图 SITE PLAN

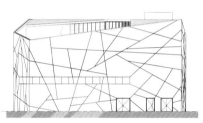
南立面图 SOUTH ELEVATION

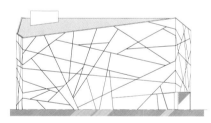
北立面图 NORTH ELEVATION

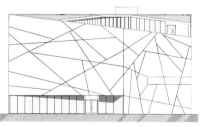
东立面图 EAST ELEVATION

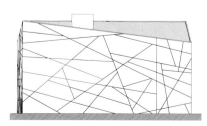
西立面图 WEST ELEVATION

© Oddleiv Apneseth

© Oddleiv Apneseth

中绿宝石 Medium Turquoise

 黄土赭色 Sienna

 纳瓦白 Navajo White

松恩&菲尤拉讷艺术博物馆，C.F. Møller Architects，挪威，弗德，2012年

进入室内，设计师大胆地在旋转的楼梯部分运用宝石绿颜色，旋转的楼梯和空间形成了鲜明的对比，增强了视觉效果。

Sogn & Fjordane Art Museum, C.F. Møller Architects, Førde, Norway, 2012

In the interior, the architect uses turquoise for the spiral stairs. The stairs present a striking contrast against the whole interior space, strengthen the visual impact.

材料：木板、磨光混凝土地板、染色混凝土、金属网顶棚

Materials: wood slats, polished concrete floors, painted concrete, metal mesh ceilings

© Stein Sandemose Baardsen

1 主入口	1 Main Entrance
2 咖啡厅	2 Café Sobra
3 招待区	3 Reception
4 衣帽间／卫生间	4 Wardrobe/WC
5 电梯（通往地下室和停车场）	5 Lift (access from undercroft parking)
6 博物馆商店	6 Museum Shop
7 主楼梯	7 Main Stairs
8 特别展会	8 Special Exhibiton
9 教学区	9 Teaching
10 工作坊	10 Workshops
11 交付区	11 Deliveries
12 员工入口／逃生楼梯	12 Staff Entrance / Escape Stairs

一层平面图 FIRST FLOOR PLAN

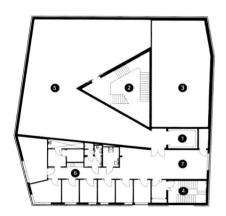

1 电梯（通往地下室和停车场）	1 Lift (access from undercroft parking)
2 主楼梯	2 Main Stairs
3 特别展会	3 Special Exhibiton
4 员工入口／逃生楼梯	4 Staff Entrance / Escape Stairs
5 展览空间 2	5 Exhibition Space 2
6 管理室	6 Administration
7 图书馆	7 Library

二层平面图 SECOND FLOOR PLAN

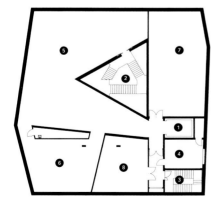

1 电梯（通往地下室和停车场）	1 Lift (access from undercroft parking)
2 主楼梯	2 Main Stairs
3 员工入口／逃生楼梯	3 Staff Entrance / Escape Stairs
4 管理室	4 Administration
5 展览空间 3	5 Exhibition Space 3
6 展览空间 4	6 Exhibition Space 4
7 临时存储区	7 Temporary Storage
8 道具室	8 Props Room

三层平面图 THIRD FLOOR PLAN

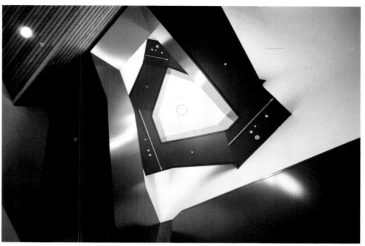

© Oddleiv Apneseth

© Stein Sandemose Baardsen

意象与现实
IMAGE REALITY

 鹿皮色 Moccasin

 秘鲁色 Peru

 重褐色 Saddle Brown

① 房屋板壁，Henrique Oliveira，法国，巴黎，瓦卢瓦画廊，2008年
Tapumes, Henrique Oliveira, Galerie Vallois, Paris, France, 2008
3.2 m × 6.2 m × 0.9 m
材料：胶合板、PVC
Materials: plywood, PVC

② 建筑突出物，Henrique Oliveira，巴西，圣保罗，巴罗克鲁兹画廊，2008年
Architectonic Intumescence, Henrique Oliveira, Baró Cruz Gallery, São Paulo, Brazil, 2008
3.4 m × 1.8 m × 1.2 m
材料：胶合板、PVC
Materials: plywood, PVC

③ Leões房屋板壁，Henrique Oliveira，巴西，阿雷格里港，南锥共同市场双年展，2009年
Tapumes - Casa dos Leões, Henrique Oliveira, VII Bienal do Mercosul, Porto Alegre, Brazil, 2009
材料：胶合板、PVC
Materials: plywood, PVC

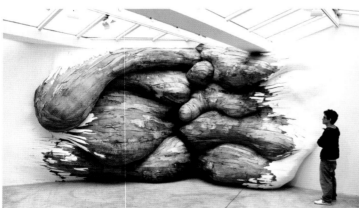
© Henrique Oliveira ①

© Henrique Oliveira ②

© Eduardo Ortega

© Mauro Restiffe ①

© Nash Baker ②

鹿皮色 Moccasin

 秘鲁色 Peru

 重褐色 Saddle Brown

 淡珊瑚色 Light Coral

① 旋风，Henrique Oliveira，巴西，圣保罗，英国大使馆，2007年
Whirlwind for Turner, Henrique Oliveira, Conselho Britânico, São Paulo, Brazil, 2007
4.35 m×6.92 m×2 m
材料：胶合板
Materials: plywood

② 房屋板壁，Henrique Oliveira，美国，休斯敦，莱斯画廊，2009年
Tapumes, Henrique Oliveira, Rice Gallery, Houston, USA, 2009
4.7 m×13.4 m×2 m
材料：胶合板、颜料
Materials: plywood, pigment

③ 房屋板壁，Henrique Oliveira，巴西，圣保罗，菲亚特视觉艺术大赛，2006年
Tapumes, Henrique Oliveira, Fiat Mostra Brasil, São Paulo, Brazil, 2006
4 m×18.4 m×1.2 m
材料：胶合板
Material: plywood

© Mauro Restiffe

■ "白有漂白、月白。" —— 清李斗《扬州画舫录·草河录上》■白色，象征纯洁、无邪、干净、完美。■众多建筑师对简朴之最的白色情有独钟，构成简洁明确、朴素有力的视觉效果。■所谓"无中万般有"的东方式虚无在白色中得到了很好的印证和体现。■著名作家川端康成曾说过："没有杂色的洁白，是最清高也是最富有色彩的。"■不管是自然界的白还是白色涂料的白，都没有纯净的白，同样在空间中的白色也是丰富的，除了自身涂料的色差，还会受到环境色的影响，因此在设计中，经过与其他色彩调和的白色更为柔和、适用，可消除纯白带来的枯燥和刺目。

■ White is the color of the moon. ■ White stands for purity, innocence, cleanness and perfection. ■ Many architects show special preference to white, the plainest color to create simple, clear, pristine and powerful visual effects. ■ The oriental nihility that "everything is contained in nothingness" finds its well confirmed and expressed in white. ■ The famous writer Kawabata Yasunari once said, "White is the cleanest of colors, it contains in itself all the other colors." ■ Neither the white color exists in nature nor in paint is pure. Likewise, the white color used in the spaces is various. Apart from its own chromatic aberration, white is influenced by the environmental colors. White color, after being fixed with other colors, will therefore become softer and more available, and remove the dullness and harshness of pure white.

月白之洁
WHITE
245

花开物语
FLOWER IMPLICATION

雪白色 Snow

 秘鲁色 Peru

© Ryuji Nakamura

盛开(餐厅室内设计), Ryuji Nakamura, 日本, 长野, 2009年

Ryuji Nakamura的这个非常有诗意的项目名为"盛开", 是位于日本长野一间餐厅的室内设计。房间漆成白色, 金属板被切割成花的形状, 同样漆成白色。这些翘曲的花被粘在一面平坦的墙上, 与墙面若即若离。花的数量约为12 000个。光影使花的轮廓浮现在墙体表面上, 浪漫之极。

Blossom (Interior Design of a Restaurant), Ryuji Nakamura, Nagano, Japan, 2009

This project "blossom" is the interior design of a restaurant in Nagano, Japan. Again Ryuji Nakamura presented with a very poetic concept. The room is painted white and a series of metal petals of the same color are attached to the flat walls. About 12,000 warping petals flirt with the flat walls. Light and shadows create a romantic floral pattern on the wall surface.

材料: 金属板、白色漆

Materials: metal panel, white paint

朦胧之美
HAZY BEAUTY

 小麦色 Wheat

 茶色 Tan

 粉蓝色 Powder Blue

© Pedro Pegenaute

© Pedro Pegenaute

巴乔马丁市政厅，Magén Arquitectos，西班牙，特鲁埃尔，2011年

建筑由半透明的雪花石膏和不透明的石灰石建造而成，材料都取自当地的采石场。石材与雪花石膏制成的切割体量通过抽象的方式和几何形态影射当地采石场的石材形态。不透明和半透明的石材表面展示了雪花石膏的材料特性与表现力。在日光或夜晚灯光下呈现不同的效果。

Bajo Martín County Seat, Magén Arquitectos, Teruel, Spain, 2011

The building is constructed from translucent alabaster and opaque limestone that were extracted from native quarries. The group of carved volumes on local materials – stone and alabaster, alludes, in an abstract and geometric way, to stone groups that occur in quarries in the area. The stone surfaces, opaque or translucent, exhibit materials and expressive features of alabaster in relation to the day or night lighting.

材料：雪花石膏、不透明的石灰石

Materials: alabaster, limestone

© Pedro Pegenaute

浅灰色 Light Gray

 灰菊黄 Pale Goldenrod

 深灰色 Dark Gray

© Pedro Pegenaute

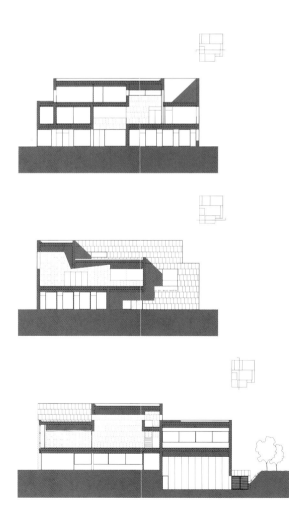
纵向剖面图 LONGITUDINAL SECTIONS

© Pedro Pegenaute

陶坯黄 Bisque

 黄土赭色 Sienna

 重褐色 Saddle Brown

巴乔马丁市政厅，Magén Arquitectos，西班牙，特鲁埃尔，2011年

外部材料在室内的连续使用以及通过各种缝隙进入室内的自然光线增强了室内的空间效果，室内就好像是一个挖空的空间。在室内，温馨的竹木饰面与粗糙的集合造型形成对比，白色的石墙影射了传统伊比利亚地方语言的沉静与朴素，同时也利用了取自当地采石场的材料。

Bajo Martín County Seat, Magén Arquitectos, Teruel, Spain, 2011

The continuity with the outside material and the presence of natural light into the interior through various gaps, strengthen the condition of the interior space as empty excavated. In the internal space, the warmer, softer bamboo finish contrasts with the harsh geometry. White stone walls allude to the sobriety and plainness of traditional Iberian vernacular as well as referencing material groups from local quarries.

材料：雪花石膏、不透明的石灰石、木材

Materials: alabaster, limestone, wood

© Pedro Pegenaute

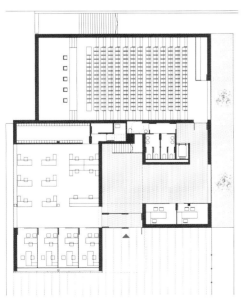
一层平面图 GROUND FLOOR PLAN

© Pedro Pegenaute

与风共舞
MOVE WITH THE WIND

 雪白色 Snow

 灰色 Gray

西立面图 WEST FAÇADE

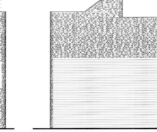
南立面图 SOUTH FAÇADE

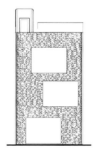
东立面图 EAST FAÇADE

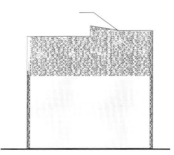
北立面图 NORTH FAÇADE

© John Lewis Marshall

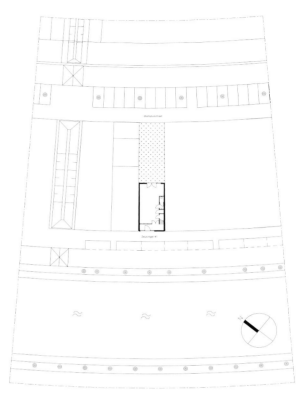
总平面图 SITE PLAN

织物立面公寓，cc-studio & studiotx & Rob Veening，荷兰，阿尔梅勒，2010年

织物立面公寓拥有独特的表皮，柔韧的材料可以随风摆动，仿佛穿着一件飘逸的白色织布衫裙。这些极为耐用的不易燃材料为带有聚四氟乙烯涂层的玻璃纤维织物，来自于食品制造业中废弃的传送带。材料经过精心裁剪，相互搭接，钉在刨花板上，构成了建筑的外墙。柔韧的材料可以随风摆动，营造出鲜活灵动的立面形象。建筑中央有一个中庭，顶部设有大天窗，视线可贯通各楼层。室内形成一个连续的空间，光线、方位都十分舒适。室内以白色为主色调，使空间宽敞、明亮，营造出清新自然的环境。

Fabric Façade: Studio Apartment, cc-studio & studiotx & Rob Veening, Almere, the Netherlands, 2010

The building has a unique skin, and the flexible material moves with the wind, like a white ethereal woven dress over the building. The extremely durable, non-combustible, residual material comes from rolls of PTFE (Teflon) coated fiber glass fabric, used in the industrial manufacture of conveyors belts for the food industry. It was cut and patterned and tacked as overlapping shingles on a backing of OSB panels. The flexible material moves with wind, creating a lively image. In the middle of the building volume a central atrium with at the top a skylight was carved out, visually connecting all the upper floors. A continuous internal space is the result which has a very pleasant light quality and orientation. The interior is in white, spacious and bright, creating a fresh and natural environment.

材料：带有聚四氟乙烯涂层的玻璃纤维织物

Material: PTFE (Teflon) coated fiber glass fabric

© John Lewis Marshall

银妆素裹
A WHITE WORLD IN SNOW

雪白色 Snow

亮蓝 Light Blue

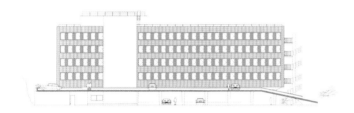

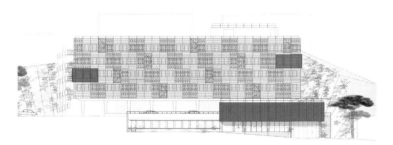

立面图 ELEVATIONS

© Courtesy of ECDM architectes

总平面图 SITE PLAN

法国电力集团总部的改造，Emmanuel Combarel Dominique Marrec architectes(ECDM)，法国，阿雅克修，2012年

在这座建筑的改造中，设计者意在打造一个简洁明了而精巧雅致的体量。建筑采用了纯粹统一的立面处理，并重新解读了业已存在的遮阳层。在建筑的南侧，横向的喷涂铝材板条完全覆盖了朝南的主立面。立面上布满了纤长的开口，这使得立面的比例富有变化，柔化了建筑的外观，并与周围的建筑和景观融为一体。板材叶片在图案上呈现出微妙的变化，为立面带来了蓬勃的生机。从远处观望，建筑就像是冰雪覆盖的雪山，白雪皑皑，在柔和的月光下映衬下，更显得皎洁明亮，充满神秘气息，让人不免心驰神往。

Restructuration du siège EDF Ajaccio, Emmanuel Combarel Dominique Marrec architectes(ECDM), Ajaccio, France, 2012

During the reconstruction, the designer attempted to present a simple and elegant building. They propose a unitary treatment of the façades, while reinterpreting the concept of sunscreen already in place. The south side is equipped with horizontal strips of lacquered aluminum, which are completely covering the main façade facing south. With the slender openings all over, the façade changes its scale and offers a softer perception of the volume, in keeping with its surroundings and the local landscape. Subtle shifts in the pattern of the blades create vibrations for the façade. The building looks like a snow mountain when seen from afar, with "snow" glowing in the soft moonlight. It is such a mysterious building that draws people's attention.

材料：喷漆铝材

Material: lacquered aluminum

© Courtesy of ECDM architectes

© Courtesy of ECDM architectes

白色时光
WHITE TIME

雪白色 Snow

 海洋绿 Sea Green

© Serge Demailly

西蒙娜公寓，Jean-Pierre Lott Architecte，摩纳哥，2012年

这座建筑旨在开启一项住宅类型学的新门类，超越建筑本身，对城市结构进行分析，努力解决城市高密度问题。这座高层建筑为其内设的公寓创造了最佳的开放视野。其非同寻常的建筑语汇使城市的色彩更加丰富。建筑优雅地屹立在群山之中，白色的外立面在周围绿植的衬托下，显得格外纯净，似天边飘落下的一抹云。而立面上富有设计感的镂空装饰，又为建筑平添了几分韵律。

The Simona, Jean-Pierre Lott Architecte, Monaco, 2012

This building seeks to deliver a study in housing typology and, beyond the building itself, an analysis of urban structure and a response to the question of the city's densification. The high-rise building provides great comfort for its apartments: the best views and an open outlook. Its unusual vocabulary will broaden the city's palette. Viewed form outside, the building stands gracefully among the hills. Contrasted by the surrounding greenery, the building is uniquely pure and clean, like a cloud fluttering down the sky. Further more, the exquisitely designed hallowed-out patterns on the façade add an extra rhythm to the building.

材料：混凝土

Material: concrete

© Serge Demailly

雪白色 Snow

 玫瑰棕色 Rosy Brown

 浅海洋绿 Light Sea Green

暗淡灰 Dim Gray

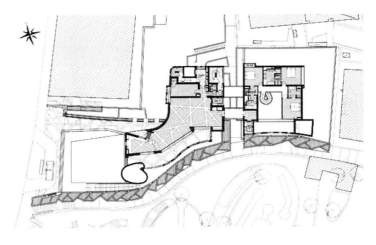

平面图 PLAN

© Serge Demailly

© Serge Demailly

© Serge Demailly

© Serge Demailly

© Serge Demailly

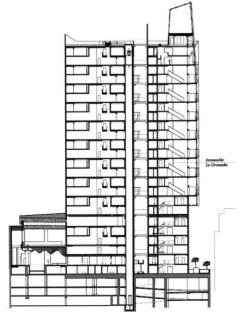
剖面图 SECTION

西蒙娜公寓，Jean-Pierre Lott Architecte，摩纳哥，2012年

白色是纯洁的象征，通过建筑结构和光线的折射，造就了光影斑驳的空间。空间和光线之间相互契合，彼此交融，令整个空间流光溢彩。

The Simona, Jean-Pierre Lott Architecte, Monaco, 2012

White stands for purity. A dappled space is created through the structure of the building and the refraction of light. Light integrates with the space and makes the space bright and colorful.

材料：混凝土、玻璃

Materials: concrete, glass

纯净的世界
THE WHITE AND PURE WORLD

雪白色 Snow

 暗黄褐色 Dark Khaki

 橄榄褐色 Olive Drab

贝纳通幼儿园，Alberto Campo Baeza，意大利，特雷维索，庞萨诺威尼托，2007年

在茵茵绿草地上，屹立着一座白色的建筑，很难想象这是一所幼儿园。建筑物通体洁白，圣洁的感觉油然而生。建筑就像是一个方盒子，由9个小的方形建筑构成。中间的方形体量从高高的顶棚上引入自然光线，通透而敞亮；教室布局在其周围。这个方形结构由更大的圆形外壳组成的双弧墙围合起来，形成了4个开敞的院落，它们代表了4种元素：空气、泥土、水和火。

Day Care Center for Benetton, Alberto Campo Baeza, Ponzano Veneto, Trevisol, Italy, 2007

A white building stands in the green meadow, which is hard to be associated with kindergarten. The building is all white, gives a saintly feeling. The building is built as a square box composed of nine smaller squares. The center square emerges to bring light from the heights of the vestibule. The classrooms are arranged in the surrounding squares. This square structure is inscribed within a larger, circular enclosure made up of double circular walls. Open to the sky, four courtyards are created that suggest the four elements: air, earth, fire and water.

© Hisao Suzuki

© Marco Zanta

雪白色 Snow

 硬木色 Burly Wood

 暗黄褐色 Dark Khaki

贝纳通幼儿园，Alberto Campo Baeza，意大利，特雷维索，庞萨诺威尼托，2007年

中央方形空间拥有屋顶采光，顶棚上的9个圆孔和每个立面上的3个圆孔为建筑提供了良好光感，给人一种古典浴场的感觉。室内为白色，利用光影的变化来赋予室内空间以生命的灵性。

Day Care Center For Benettont, Alberto Campo Baeza, Ponzano Veneto, Trevisol, Italy, 2007

The central space, the highest and with light from above, recalls a hamman in the way it gathers sunlight through nine perforations in the ceiling and three more on each of its four façades. The interior is in white, and the gradations of light and shade bring the spiritualism of life to space.

© Marco Zanta

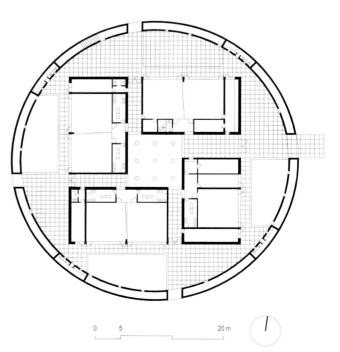

平面图 PLAN

© Marco Zanta

© Marco Zanta

涟漪轻漾
GENTLE RIPPLES ON THE SURFACE

 雪白色 Snow

 银灰色 Silver

 灰色 Gray

总平面图 SITE PLAN

© Benoît Fougeirol

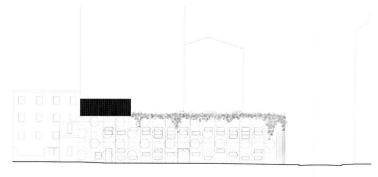

立面图 ELEVATION

皮埃尔·布丁街托儿所,Emmanuel Combarel Dominique Marrec architectes (ECDM),法国,巴黎,2012年

建筑的外墙由白色预制混凝土面板构成,通体洁白,无比纯净。立面形成一种波浪起伏的效果,宛如水面上的涟漪一般。外墙只有一侧装有透明的彩色窗户,其高度不一,能够满足不同人群的视野要求。在这里,无论是大人还是小孩,无论是父母还是工作人员,都能感到温馨与关怀。

Crèche Rue Pierre Budin, Emmanuel Combarel Dominique Marrec architectes (ECDM), Paris, France, 2012

The wall of the building is made of white prefabricated concrete panels. It's white in the entire body, clean and pure. The façade creates an undulating effect. It's like ripples on the surface of water. The surrounding wall is drilled by translucent and colored windows. These windows have various heights, for a place thought as much for the children than for the adults, the parents or the staff.

© Benoît Fougeirol

材料:混凝土

Material: concrete

灰菊黄 Pale Goldenrod

 亮蓝 Light Blue

 银灰色 Silver

暗淡灰 Dim Gray

© Benoît Fougeirol

© Benoît Fougeirol

皮埃尔·布丁街托儿所，Emmanuel Combarel Dominique Marrec architectes (ECDM)，法国，巴黎，2012年

室内空间洁白，明亮，给人一尘不染的素雅感觉。干净的色调与简洁的设计诠释出空间的雅致与柔美。这里的孩子们就像是一群美丽洁白的小天使，拥有纯洁无瑕的灵魂。

Crèche Rue Pierre Budin, Emmanuel Combarel Dominique Marrec architectes (ECDM), Paris, France, 2012
The interior is pure white and bright, giving people a sense of simpleness and elegance. The exquisiteness and delicacy of the space are perfectly expressed by the clean colors and the simple design. The children are like beautiful angels, with clean and pure soul.

材料：混凝土、玻璃

Materials: concrete, glass

© Luc Boegly

© ECDM

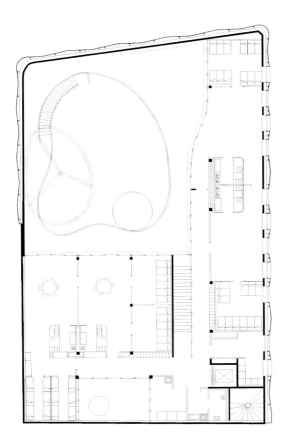
平面图 PLAN

优雅·通透·轻盈
GRACEFUL CRYSTAL ETHEREAL

 淡青色 Light Cyan

纯黄 Yellow

 浅灰色 Light Gray

暗海洋绿 Dark Sea Green

路易布兰科大街45号社会住宅，Emmanuel Combarel Dominique Marrec architectes (ECDM)，法国，巴黎，2006年

朝向街道的北立面由窗口和蛋白石保温板构成，突出垂直设计，为与其达成平衡，相辅相成，朝向庭院的南立面采用不同的设计，由一系列错落的楼层构成。该建筑坐落在一条不起眼的街道上，温和、轻盈，却极具存在感，宛如传统、厚重的建筑环境中的一颗优雅、璀璨的宝石。

Social Housing in 45 Rue Louis Blanc, Emmanuel Combarel Dominique Marrec architectes (ECDM), Paris, France, 2006

The north façade – with its windows and "opalite" insulation panels, which faces the street and emphasizes verticality – is counterbalanced by the rear of the building that faces the courtyard and has southern exposure, and has been designed differently – with successively receding stories. The building has the effect of a jewel of gracefulness set in a conventional, heavy architectural environment: a building of non-aggressive, ethereal appearance with an uncommon strength of presence implanted in a nondescript street.

材料：蛋白石石板

Material: opalite panels

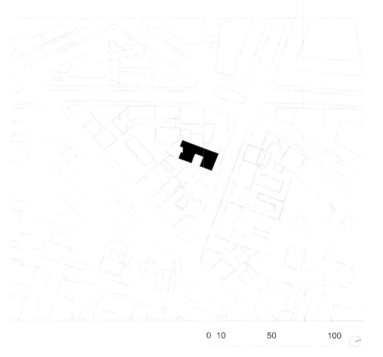

0 10   50   100

总平面图 SITE PLAN

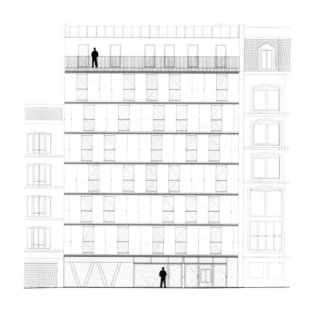

0 1   5   10

立面图 ELEVATION

© Benoit Fougeirol

© Benoît Fougeirol

在云端
IN THE CLOUDS

雪白色 Snow

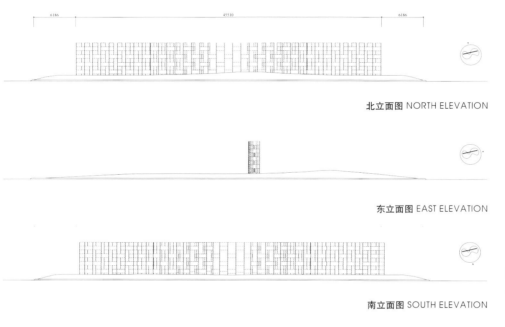

北立面图 NORTH ELEVATION

东立面图 EAST ELEVATION

南立面图 SOUTH ELEVATION

西立面图 WEST ELEVATION

陶瓷云，Kengo Kuma & Associates，意大利，卡萨尔格兰德，2010年

建筑精妙的半透明性和每块瓷砖表面的反光性能够因天气状况和一天的时间段而变化万千。建筑在白天和夜晚的视觉效果大相径庭。在日光的沐浴下，瓷砖与这一繁华的区域、周围环境和地中海气候氛围和缓地融为一体。纯白色的建筑安静地矗立，似站在云端的天使，守护这一方净土。

CCCloud, Kengo Kuma & Associates, Casalgrande, Italy, 2010

The subtle translucency of the monument and reflections that appear on the surfaces of each tile allows it to interact with the weather and time of day. The monument's day and night effects vary substantially. In the light of day the tiles blend serenely with the developed site, the surrounding area, and the atmosphere of the Mediterranean climate. The pure white building stand silently like an angle in the clouds, who has been guarding the pure land.

© Marco Introini

材料：瓷砖

Material: ceramic tiles

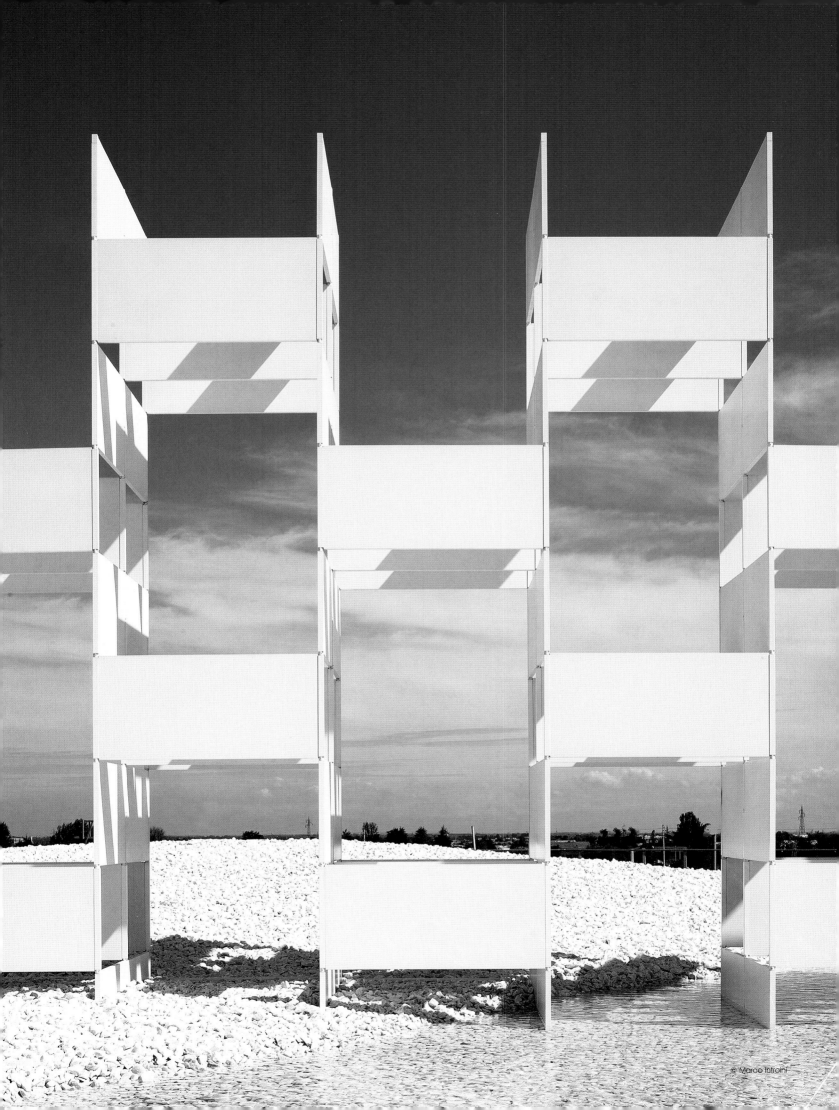

雪白色 Snow

© Marco Introini

陶瓷云，Kengo Kuma & Associates，意大利，卡萨尔格兰德，2010年

当夜幕降临时，建筑所处环境笼罩在黑暗之中，建筑则凭借其反光性亮洁而动感地傲然耸立。这一装置在视觉上绝不是静止不动的。在这个简洁至极却诗情画意的装置中，光、水和瓷砖互动交叠，不断交织成迷人的刹那芳华。它的存在为这一地区带来了宁静感和更为强烈的辨识度。

CCCloud, Kengo Kuma & Associates, Casalgrande, Italy, 2010

At night, as the background fades into the darkness, the CCCloud stands illuminated and vibrant in its reflection. The installation is by no means visually static. The effect of the ceramic tiles creates an ephemeral play of light, water and material in the minimal and poetic installation. The monument evokes tranquility and mindfulness of the site through its presence.

材料：瓷砖

Material: ceramic tiles

© Marco Introini

轻盈之美
THE BEAUTY OF LIGHTNESS

雪白色 Snow

灰菊黄 Pale Goldenrod

重褐色 Saddle Brown

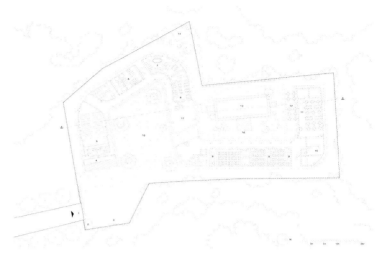

1 主入口	1 Main Entrance
2 监护室	2 Protect Room
3 教师车库	3 Garage for Teacher
4 屋顶入口	4 Entrance to Roof
5 多功能厅	5 Multipurpose Hall
6 办公室	6 Office
7 会议室	7 Meeting Room
8 学生阅览室	8 Reading Room for Pupils
9 教室	9 Classroom
10 厨房	10 Kitchen
11 食堂	11 Canteen
12 露台	12 Terrace
13 游泳池	13 Swimming
14 工程部门	14 Engineering Department
15 公共庭院	15 Public Court
16 学生庭院	16 Student Court
17 公共空间	17 Common Space

总平面图 SITE PLAN

© Hiroyuki Oki

© Hiroyuki Oki

平阳学校，Vo Trong Nghia Architects，越南，平阳，2011年

项目的建筑场地位于一片繁茂的森林之中，绿植和水果树葱葱茏茏。建筑斑驳的倒影与周围的树影相映成趣，白色的外立面使整个建筑看上去轻盈柔美。这座建筑位于5 300 m²的广阔土地上，最高处的楼层为5层，四周被森林所包围。建筑的表皮由预制混凝土栅格和带有图案的墙壁构成。这些遮阳装置成就了建筑半户外型的空间，避免阳光的直射，并充当走廊的自然通风系统。所有的教室均以这个半开放空间为连接纽带，为老师与学生聊天、交流和欣赏自然美景提供了场所。

Binh Duong School, Vo Trong Nghia Architects, Binh Duong, Vietnam, 2011

The site is located in the middle of a flourishing forest with a wide variety of green and fruits, running rampant. The mottled shadows of the building contrast finely with the surrounding tree shadows. The white façade adds light and soft beauty to the entire building. The building is located in 5,300 square meters abundant land, consisting of a maximum height of five levels, with the intention of being surrounded by the height of the forest around. Precast concrete louvers and pattern walls are used for envelop of the building. These shading devices generate semi-outside space, these open circumstances avoiding direct sunlight as well as acting like a natural ventilation system for the corridor space. All the classrooms are connected by this semi-open space, where teachers and students chatting, communicating and appreciating nature.

材料：预制混凝土

Material: precast concrete

© Hiroyuki Oki

光的游乐场
HOUSE OF LIGHT

 雪白色 Snow

 秘鲁色 Peru

 深灰色 Dark Gray

 黄土赭色 Sienna

 天窗 House of Light (HOL)

平面图 PLAN

雷蒙德幼儿园，Archivision Hirotani Studio，日本，滋贺县，长滨，2011年

　　光色相随，光强色强，幼儿深谙此道。因此建筑师设计了"光之家"幼儿园，用光与色为幼儿创造出一片乐园。幼儿园采用单层的结构，各种不同风格的窗型设计贯穿其中，将光线和室外的美景引入。光赋予其更独特的性格，使幼儿园成为一座名副其实的"光之家"。建筑师称幼儿园为"光之家"，此说法源自于建物利用屋顶空间的锥形设计，各个形状的锥块结构，形成数个"光线引入装置"，将来自不同方位的自然天光引入游戏区、卧房、浴室之中，反映时间和季节的变换，试图让孩子在生活与玩乐中享受阳光的洗礼。孩子们能感觉到这些"灯"的变化，甚至追逐他们并一同玩耍，在日常生活以及活动中享受这光的礼物。同时，"光之家"的外形轮廓为这一成不变的田园风光带来些许惊喜。

The Leimond Nursery School, Archivision Hirotani Studio, Nagahama, Shiga Prefecture, Japan, 2011

Children are sensitive with colors and light. With this in mind, the architects designed this nursery school, creating a fairyland for children with colors and light. The nursery school has been planned as a single-story structure with a feeling of transparency between each of the spaces as well as the exterior landscape and, the "House of Light", as we call it, has been placed in the main nursery area. What the architects mean by the "House of Light" are conical, square light-wells of different shapes, different color and facing different directions in the high ceiling bringing in various "lights" into the interior space, changing with the time and the seasons. The children may be able to feel the changes of these "lights", even chase them and play with them, and to enjoy this gift of "light" in their daily activities. Furthermore, the shape of the "House of Light" may be seen from the outside as its unique silhouettes are outlined against the almost unchanging rural scenery, providing it with a little more character.

材料：木材

Material: wood

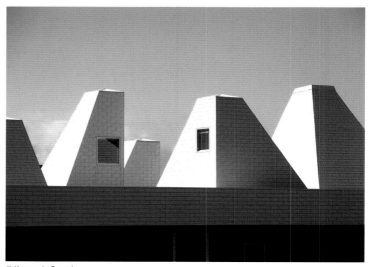

© Kurumata Tamotsu

© Kurumata Tamotsu

© Kurumata Tamotsu

© Kurumata Tamotsu

© Kurumata Tamotsu

硬木色 Burly Wood

亮粉红 Light Pink

纯黄 Yellow

绿黄色 Green Yellow

黛青色 Dark Cyan

Leimond幼儿园，Archivision Hirotani Studio，日本，滋贺县，长滨，2011年

在幼儿园的设计中有两个重要因素：光与色彩。色彩对幼儿是一种不可抵挡的诱惑，他们喜欢鲜艳、明快的色彩。因此幼儿园室内主要采用如红、黄、蓝、绿、粉等色彩，为幼儿营造出五彩斑斓的世界，使孩子们仿佛置身于童话王国。

The Leimond Nursery School, Archivision Hirotani Studio, Nagahama, Shiga Prefecture, Japan, 2011

Light and color are two key elements in kindergarten design. Color is an irresistible temptation for children. They like bright and lively colors, so the interior of this kindergarten applies striking colors, such as red, yellow, blue, green, pink, etc., creating a colorful world for the children. It's like a fairy kingdom for them.

材料：木材、涂料

Materials: wood, paint

剖面图 SECTION

© Kurumata Tamotsu

流动的音符
FLOWING MUSIC

雪白色 Snow

秘鲁色 Peru

鹿皮色 Moccasin

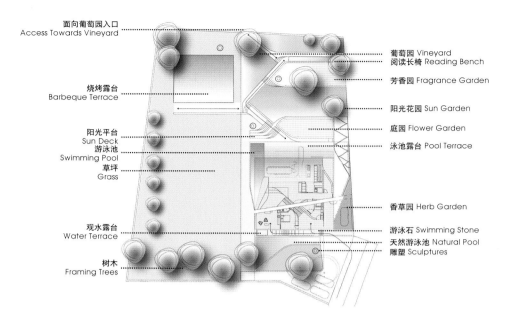

景观设计分析图 LANDSCAPE DESIGN DIAGRAM

温伯格别墅，UNStudio，德国，斯图加特，2008年

　　温伯格别墅的体量和屋顶轮廓线与该地倾斜的景观产生了直接的互动和呼应，白色的外立面在绿色葡萄园的映衬下格外醒目。斜坡雕刻出了葡萄园的背景，其比例和倾斜度在建筑体量感十足的外观上得以体现。设计师对建筑的内部流通路线、视野的安排，以及空间的布局采用一种"扭转"的姿态。建筑空间布局顺应了太阳的运动轨迹，同时每处扭曲都使户外的视野纳入了室内，身处其中，让人分不出室内还是室外。室内装有天然橡木地板、天然石材和嵌着晶莹石片的白灰泥墙，其内部材质进一步强化了别墅整体的光线氛围。特别定制的装置和家具与建筑融为一体，并突出了建筑特色。

Haus am Weinberg, UNStudio, Stuttgart, Germany, 2008

The volume and roofline of the house react and respond directly to the sloping landscape of the site, where the scales and inclinations of the slopes which sculpture the vineyard setting are reflected in the volumetric appearance of the house. Set off by the green vineyard, the white façade becomes more eye-striking. The inner circulation, organization of the views and the programme distribution of the house are determined by a single gesture, "the twist". Whilst the programme distribution follows the path of the sun, each evolution in the twist leads to moments in which views to the outside become an integral experience of the interior. It is difficult to distinguish exterior fromm interior. The materialization of the interior of the house further accentuates the overall atmosphere of light by means of natural oak flooring, natural stone and white stucco walls speckled with small fragments of reflective stone. Custom made features and furnishings are also integrated to blend with and accentuate the architecture.

© Iwan Baan

© Iwan Baan

材料： 室外 钢筋混凝土、铝、玻璃
室内 木材、石材、白灰泥

Materials: Exterior: reinforced concrete, aluninum, glass
Interior: timber, stone, white stucco

© Iwan Baan

亲近自然
CLOSE TO NATURE

 雪白色 Snow

 深灰色 Dark Gray

 纯红 Red

博索莱伊小学，Calori Azimi Botineau Architects，法国，博索莱伊，2009年

建筑师参照天然地形的变化，将建筑的外形设计成当地的传统梯田风貌，仿佛隐嵌于地势之中。建筑总体为钢筋混凝土结构。楼面板加厚的边缘采用了凿石锤雕凿，而包含门卫室的塔形部分为混凝土结构，它经过空气钻的冲击，将骨料暴露在外。

Primary School in Beausoleil, Calori Azimi Botineau Architects, Beausoleil, France, 2009

The profile of the building follows the slope of the natural terrain and adheres to the view servitude. It is anchored and begins to disappear into the topography. The ensemble of the structure of the project is in reinforced concrete. The rim forming the thickness of the floor plates is bush hammered and the small tower comprising the wardens house is concrete, shattered, by a Pneumatic drill to expose the aggregate.

材料：钢筋混凝土、钢材

Materials: reinforced concrete, steel

© Serge Demailly

© Serge Demailly

© Serge Demailly

© Serge Demailly

银灰色 Silver

 军兰色 Cadet Blue

 纯黄 Yellow

 橙色 Orange

 火砖色 Fire Brick

© Serge Demailly

© Serge Demailly

博索莱伊某小学，Calori Azimi Botineau Architects，法国，博索莱伊，2009年

　　学校室内色彩亮丽，与外部的灰白色调形成互补。黄色、粉色等清新亮丽的色彩，带给人们阳光般的温暖，让人感觉温馨舒适，心情愉悦。

Primary School in Beausoleil, Calori Azimi Botineau Architects, Beausoleil, France, 2009

The interior is in bright colors, complementing to the gray tone in the exterior. The fresh and bright colors such as yellow and pink bring warm feelings like sunshine, making people feel very comfortable and pleasant.

材料：钢筋混凝土、钢材

Materials: reinforced concrete, steel

■蓝，染青草也。——《说文》■蓝色，颜色中的贵族。英国贵族被称为"蓝血"，皇室和王族女性所穿的深蓝色服装的颜色被称为"皇室蓝"。蓝色，亦是颜色中的明星，以蓝色命名的音乐、书籍、明星、珠宝举不胜举。■纯净的蓝色表现出一种美丽、冷静、理智、安详与广阔，是永恒和力量的象征。蓝色还代表希望，因此有"蓝图"一说。■蓝色更容易与天空或海洋相协和，大面积的玻璃面积反射天空而让整个建筑呈现蓝色的设计一度风靡。■在室内，大面积的深蓝色则显得寒冷凄凉，使用要慎重。暖蓝派生色适合墙壁、顶棚和地板。另外，加入大量白色的浅蓝也可以用于大多数的空间。

■ Blue originally referred to indigo plant. ■ Blue is a noble among colors. British aristocrats are called "blue blood", and the dark blue of clothes worn by female royals is called "royal blue". Blue is also a superstar among colors. The music, books, stars, jewelries named after blue are numerous to list. ■ Pure blue presents a sort of beauty, calm, reason, serenity and broadness. It is the symbol of eternity and power. Blue stands for hope, hence the word "blueprint" refers to the future. ■ Blue is easier to harmonize with the sky or the sea. It was all the rage to use glass on architecture façades, tinting the building with blue by reflecting the color of the sky. ■ If used in interior, large scale of dark blue will create a cold and dreary feeling, so it shall be used properly. Warm derivative colors of blue can apply to walls, ceilings and floors. Besides, whitish blue is also suitable to most spaces.

蓝之娴静
BLUE
287

精灵之家
ELF HOME

 深天蓝 Deep Sky Blue

 纯红 Red

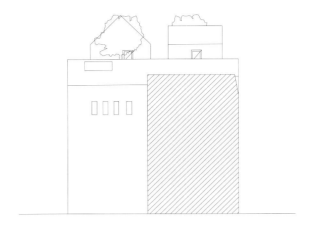

立面图 ELEVATIONS

© Rob't Hart

© Rob't Hart

迪登家园，MVRDV，荷兰，鹿特丹，2006年

这是一座屋顶上扩建的住宅。整个房子外部都被涂上了天蓝色的聚亚胺酯漆，好似打造出了一个全新的蓝色"天堂"。建筑师在院子里布置了蓝色桌子、蓝色长椅和蓝色露天淋浴，还种了树，像一个蓝色的"村庄"。这个"村庄"还被有窗户的蓝色"院墙"围绕，人们可以通过窗户观赏外面的景色。

Didden Village, MVRDV, Rotterdam, the Netherlands, 2006

This is a rooftop house extension. All the elements were finished with a blue polyurethane coating and the building looks like a crown. As a result, a new "heaven" appears. Parapet walls with windows surround the new village so that people can enjoy the street landscape through the windows. Trees, tables, open-air showers and benches are added, optimizing the rooftop life.

材料：蓝色聚亚胺酯漆

Material: blue polyurethane paint

硬木色 Burly Wood

深橙色 Dark Orange

巧克力色 Chocolate

猩红 Crimson

深天蓝 Deep Sky Blue

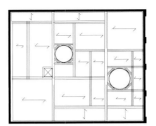

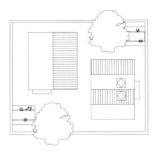

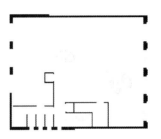

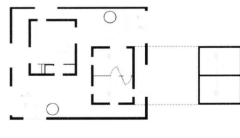

平面图 PLANS

© Rob't Hart

© Rob't Hart

© Rob't Hart

迪登村庄，MVRDV，荷兰，鹿特丹，2006年

室内的暖色调与外立面的颜色形成鲜明的对比。暖洋洋的红色与橙色搭配木质地板，纯白色的楼梯盘旋而上，暖黄色的灯光从屋顶扩散开来，温馨与浪漫的"天空之家"就这样诞生了。螺旋式楼梯的设计是这个"小村庄"的室内特色，满足了每个卧室与地面层的垂直交通。轻型木结构的螺旋式楼梯悬挂在固定于原屋顶的金属框架上，每个楼梯都通到下层的工作室，起始踏步抬升于楼地面之上，与楼地面保持了距离，像一个个"悬浮楼梯"。主卧室里的是单螺旋楼梯，而两个小卧室的楼梯相互盘旋，像DNA分子的双螺旋形结构。

Didden Village, MVRDV, Rotterdam, the Netherlands, 2006

The warm color in the interior contrasts with the color of the exterior; red and orange are chosen for the wooden flooring; the pure white stairs spiral up; the yellow light diffuse from the roof…thus a sweet and romantic "sky house" is created. The interior of the "village" features the spiral stairs, which provide vertical circulation from each bedroom to the ground floor. The wooden stairs are fixed onto the metal framework of the existing roof and each staircase leads to the studio on the lower levels and the entrance ramp was elevated above the ground, which making several suspended staircases. The staircase to the master's bedroom is single helix and two spiral stairs to the two children's houses coil around each other to form a double helix stairs.

材料：木材、金属框架

Materials: wood, metal frame

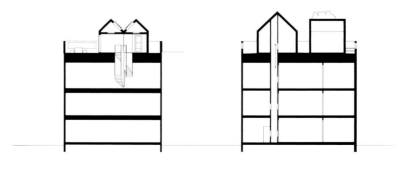

剖面图 SECTIONS

悦动流转的五线谱
FLOWING STAVE

 银灰色 Silver

 黄土赭色 Sienna

 中宝石碧绿 Medium Aquamarine

积分别墅，Shim-Sutcliffe Architects，加拿大，多伦多，2009 年

积分别墅位于多伦多峡谷地带，建筑的两层使用截然不同的材质。宝石碧绿色的玻璃清新亮丽，与木条那优美的曲线式结构相辅相成，谱写出华丽的乐章。底层由橡树条拼接作为支撑及维护，木条有疏有密，镂空的地方将玻璃嵌入其中。上层结构的支撑藏在内侧，外部由突出表面的U 形玻璃包裹，与底层木质结构在外观上分离，产生漂浮灵动的效果。上下两层外立面的轮廓都以曲线为主，如同盘旋错落的等高线，与河谷波澜起伏的岸线以及树林中蜿蜒的小径相呼应。

The Integral House, Shim-Sutcliffe Architects, Toronto, Canada, 2009

The Integral House is located at the edge of a Toronto ravine. From the street, one reads a two-story building with a grounded wood base sitting on top of translucent gently shaped etched glass skin. The turquoise glass is refreshing and clear, paired with the beautiful curved structure, composing a splendid movement. The wooden base is composed of solid walls clad in oak that dissolve into oak clad fins. The upper supporting structure is hidden inside and wrapped by a U-shaped glass, as if separating from the wood base, creating a floating effect. These serpentine walls made of vertical glass separated by projecting oak fins form the gentle perimeter to the house echoing the undulating contour lines of the river valley and the winding pathways in the native forest of oaks, maples and beeches.

材料：橡木、玻璃

Materials: oak, glass

© Ed Burtynsky

© Jim Dow

© Jim Dow

海军蓝 Navy

闪兰色 Dodger Blue

番茄红 Tomato

1 喷泉
2 游泳池
3 露台
4 健身房
5 卫生间
6 更衣室
7 淋浴/蒸汽房
8 桑拿浴
9 机械室
10 电梯

1 Fountain
2 Pool
3 Terrace
4 Exercise Room
5 Water Closet
6 Change Room
7 Shower / Steam Room
8 Sauna
9 Mechanical Room
10 Elevator

平面图 PLAN

积分别墅，Shim-Sutcliffe Architects，加拿大，多伦多，2009 年

项目进行了多方面的实验和探索，如壁炉、楼梯和门把手。其中，一个极为特别的融合性元素是蓝色的玻璃楼梯，这是一件为场地量身定制的艺术品，由玻璃艺术家Mimi Gellman、Shim-Sutcliffe Architects建筑事务所和结构工程师David Bowick合作完成。它由人工吹制的蓝色矩形夹层玻璃构成，用铸造青铜夹子和不锈钢绳固定在一起。半透明的夹层玻璃踏板贯穿激光切割的钢踏板，使得自然光线从宽大的天窗射入，通过蓝色的楼梯和玻璃踏板进入楼梯间和下方的空间。积分别墅中这座蓝色楼梯是艺术家、建筑师和工程师持续合作的产物，创造出一个缥缈的居住空间。

The Integral House, Shim-Sutcliffe Architects, Toronto, Canada, 2009

Throughout the project are many experiments and explorations – fireplaces, staircases, door handles. One special integrated element is a blue glass stair which is a site specific commissioned art work and the result of collaboration between glass artists Mimi Gellman, Shim-Sutcliffe Architects and structural engineer David Bowick is composed of hand blown laminated blue glass rectangles supported by cast bronze clips and stainless steel cables. Laminated translucent glass treads span laser cut steel treads and allow natural light to filter from a large upper skylight through the blue stair and through its glass treads to the staircase and spaces below. The Integral House's blue glass stair is the result of an on-going collaboration between an artist, architect and engineer working together to realize an ethereal space for inhabitation.

材料：蓝色玻璃、钢材

Materials: blue glass, steel

© Jim Dow

© Ed Burtynsky

世外桃源
A FICTITIOUS LAND OF PEACE

 午夜蓝 Midnight Blue

 橙红色 Orange Red

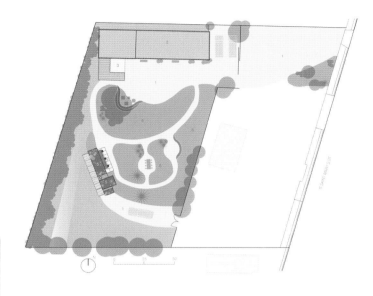

1 车道　　　1 Drive
2 阁楼住宅　2 Loft Residence
3 带纱窗的入口　3 Screened Entry
4 集装箱　　4 Container
5 屋顶绿化　5 Roof Planters
6 花园　　　6 Garden

总平面图 SITE PLAN

© Lluis Casals

© Lluis Casals

集装箱客房，Poteet Architects，美国，得克萨斯州，圣安东尼奥，2010年

设计师选用蓝色的箱体，配上橙色的家具，置之在绿树林中。蓝色象征稳重与理智，橙色代表了欢快与收获，绿色代表了舒适与希望。这个项目源于客户的一个小小愿望，他想用货船的集装箱做客房。建好的客房功能全备，巨大的玻璃和钢铁滑动结构以及尾端的一整面窗户将室内向周围的景观敞开。建筑的背面装有丝网面板，将来会爬满常青藤植物。

Container Guest House, Poteet Architects, San Antonio, Texas, USA, 2010

The architect placed a blue container equipped with orange furniture in the green forest. Blue stands for calmness and reason, orange stands for happiness and harvest while green stands for comfort and hope. This project originated in the client's wish to construct a guest house with shipping containers. The large steel and glass lift/slide and end window wall open the interior to the surrounding landscape. The rear of the container is screened by wire mesh panels which will eventually be covered in evergreen vines.

材料：集装箱、玻璃、丝网面板

Materials: container, glass, wire mesh panels

© Lluis Casals

© Lluis Casals

© Lluis Casals

© Lluis Casals

猩红 Crimson

 秘鲁色 Peru

 深天蓝 Deep Sky Blue

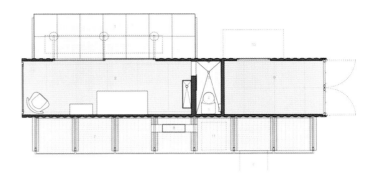

1 露天平台	1 Deck
2 工作室	2 Studio
3 浴室	3 Shower
4 顶部天窗	4 Skylight Above
5 堆肥厕所	5 Composting Toilet
6 堆肥存储池	6 Composting Bed
7 网筛	7 Mesh Screening
8 暖通空调	8 HVAC
9 储藏室	9 Storage
10 悬臂式绿植	10 Cantilevered Planters
11 灰水收集罐	11 Graywater Collection Tank

平面图 PLAN

集装箱客房，Poteet Architects，美国，得克萨斯州，圣安东尼奥，2010年

室内的空间是原木地板，红色的卫生间，呈现万绿丛中一点红的效果。住在这个山花烂漫的地方，仿佛与世隔绝，能进入到自己的世外桃源是多么美妙的一件事。建筑内部和地板铺装了喷涂泡沫和竹胶合地板来进行隔热。水槽和浴室流出的灰水可以灌溉屋顶的植物。堆肥厕所又为植物提供了肥料。

Container Guest House, Poteet Architects, San Antonio, Texas, USA, 2010

The red restroom contrasts with the wooden flooring, adding some brightness to the surrounding greenery. Living in such a place overspread with flowers is like withdrawing from the outside world. What a wonderful thing it is to live in your own fancy world. The interior is insulated with spray foam then lined with bamboo plywood, equally appropriate for the floor as the walls. The gray water from the sink and shower is captured for roof irrigation. The WC is a composting toilet.

材料：木材

Material: wood

© Lluis Casals

温馨的回忆
SWEET MEMORIES

 绿宝石 Turquoise

 军兰色 Cadet Blue

 硬木色 Burly Wood

东立面图 EAST ELEVATION

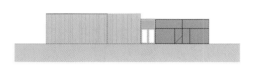

西立面图 WEST ELEVATION

南立面图 SOUTH ELEVATION

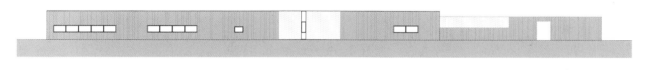

北立面图 NORTH ELEVATION

© Lluis Casals

圣伊莎贝尔幼儿园，Carroquino Finner Architects，西班牙，萨拉戈萨，2007年

室内外色调统一，宝石绿色的玻璃搭配古朴的木色，让建筑看上去灵动清雅，又有一种家的温馨。木材在建筑的构成中起到了决定性作用。这不仅仅是因为木材在室内和室外的各种框架和地板中起到了多种功用，还因为混凝土墙壁上的木质百叶纹理画出了记忆与缺失。

Nursery School Santa Isabel, Carroquino Finner Architects, Zaragoza, Spain, 2007

The color schemes of interior and exterior are unified. The turquoise glass and pristine wood provide vigor, grace and family warmth to the building. Wood plays a determining role in the compositional play, not only because of the different uses given to frames and floors both indoors and out, but also in the memory and absence sketched on the texture of the concrete walls made from the timber shuttering.

材料：混凝土、玻璃、木材

Materials: concrete, glass, wood

流动变幻的光芒之旅
THE JOURNEY OF FLOWING LIGHT

 闪兰色 Dodger Blue

 皇家蓝 Royal Blue

 钢蓝 Steel Blue

 亮菊黄 Light Goldenrod Yellow

天安Galleria百货商场，UNStudio，韩国，天安，2010年

这座建筑的主题源于商场内外的动态人流，利用外观不断变幻来吸引游客的注意。夜晚，柔和的浅色灯光在外立面形成起伏的多彩波浪。为了使用最少的照明灯具，减少溢散光，达到最大的多媒体展示面积，灯具经过特殊设计和定制生产，安置于外层铝型材竖框中。这样从外面看不见这些灯具，但是这些灯具能把光透过竖框投射到内层复合铝板上。在夜晚，灯光的投射让背墙的竖框消失不见，视错觉会无处不在。从远处看，大楼表面上的图案清晰可辨，但是近距离看，图案就会变得模糊。不同颜色的灯光为大楼披上了神秘的面纱，整座大楼散发着梦幻般的光芒。

Galleria Centercity, UNStudio, Cheonan, South Korea, 2010

The main architectural theme is originated from the dynamic flow of people both inside and outside supermarket. Moiré effects, special lighting and animations ensure that the façade changes constantly, so as to attract people eyes. Optical illusion was employed on the appearance of the building. The façade features two layers of aluminum extrusion profiles on top of a layer of composite aluminum cladding. The vertical profiles of the top layer are straight while those of the back layer are angled. This results in a wave-like appearance, which changes with the viewpoint of the spectator (Moiré effect). During the day the building has a monochrome reflective appearance. Its dynamic, undulating, silver façade corresponds with the blue sky and glows with charming light.

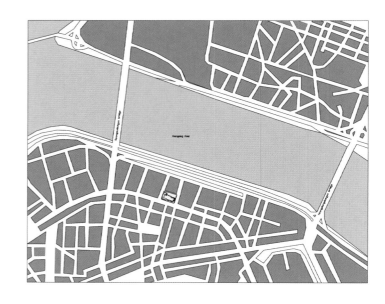

总平面图 SITE PLAN

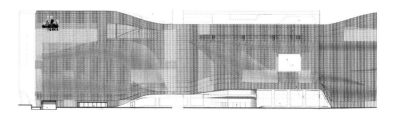

西南立面图 SOUTHWEST ELEVATION

材料：玻璃、金属型材、LED灯

Materials: glass, metal profiles, LED lights

© Kim Yong-kwan

© Kim Yong-kwan

© Kim Yong-kwan

© Kim Yong-kwan

© Kim Yong-kwan

© Kim Yong-kwan

蓝之娴静 | BLUE

天边的尖角城堡
REMOTE CASTLE WITH SHARP ANGLES

 银灰色 Silver

 淡绿色 Light Green

 亮天蓝色 Light Sky Blue

总平面图 SITE PLAN

© Koichi Torimura

街角住宅，EASTERN design office，日本，滋贺，2011年

街角住宅是"街角交叉的蓝色"，是一座由方形构件组合起来的三角形建筑。建筑的立面巧妙地利用了三角形的地块特征,仿佛是卡片散落在了外墙上。它们都被一"十字架"结合在一起，以使延伸出去的材料不会彼此分离。这座建筑看起来像是一件礼物，一个玩具箱或是童话作家迈克尔·恩德故事中孩子们可以进入的城堡。设计师力求突出剥离于城市框架的形状，它的结构更像一种错觉，如果它能"欺骗"人的视觉，让人感觉身处另一个世界，那么它便是超凡脱俗而富有超现实主义色彩的。

On the Corner, EASTERN design office, Shiga, Japan, 2011

The building is the "crossed blue on the street corner". It is a triangular building configured by the square elements. The elevation takes the shape of the triangle plot. The exterior wall is made out of square cut stone, concrete and glass formed like scattered cards on it. They are bound by a "cross" so that the spread out material would not disjoin. It looks like a present, a toy box or a castle where the boys and girls of the story of Michael Ende could be entering. The designers want to emphasize its shape from the urban framework by. The structure is more of an illusion. It seems as if this illusion deceives people to obscure their eyesight and feel invited to another world. It is pretentious, yet it is surrealistic too.

材料：石材、混凝土、玻璃

Materials: stone, concrete, glass

© Koichi Torimura

发光的树
TREES OF LIGHT

 烟灰色 Smoke Gray

 亮蓝灰 Light Slate Gray

 弱绿宝石 Pale Turquoise

Tschuggen Bergoase温泉中心，Mario Botta Architetto，瑞士，阿罗萨，2006年

温泉中心坐落在瑞士的阿罗萨地区的天然盆地中，享受着四面环山的得天独厚的地理条件。这种环境给人们带来巨大的视觉冲击力，而且尊重了村庄的原始格局，并与周围的环境和谐统一。建筑的突出之处在于其屋顶结构，建筑"伸出"9个发光的结构，模拟着山顶上的树木。另外，这些由玻璃和钢铁构成"帆"形的结构，就像是天窗，使自然光线射入建筑内部。夜晚，这些结构发出魔幻般的光线，看起来既像是发光的"叶"状雕塑，又像是白雪皑皑的群山之巅。

Tschuggen Bergoase Wellness Center, Mario Botta Architetto, Arosa, Switzerland, 2006

Arosa offers an extraordinary geographic configuration of natural basin surrounded by mountains. This particular context led to an intriguing solution of visual impact and of great respect for the surrounding village. The highlight of the design is the roof structure, where nine 'light trees' protrude from the building and point through the treetops on the mountain. These sail-shaped structures made of glass and steel, act as skylights, drawing natural light into the interior of the inset building. At night the structures are illuminated to create a magical glow, resembling luminescent leaf sculptures or snowcapped mountain peaks.

材料：钢架结构、玻璃

Materials: steel structure, glass

© Urs Homberger

© Urs Homberger

Tschuggen Bergoase温泉中心，Mario Botta Architetto，瑞士，阿罗萨，2006年

水蓝色玻璃和白色花岗岩铺砌成的天桥连接着那座20世纪60年代落成的五星级度假酒店和这个属于新世纪的温泉建筑三楼的接待处。如果从一楼直接进入，一条深沉的花岗岩楼梯同样会把你带引到三楼。"欧亚瑟冰山"各个区域之间的相互关系和与外部环境的特殊关系保证了建筑的自然采光和俯瞰外部景观的绝佳视野。在夜晚，人工照明又为建筑创造出了神秘的氛围。

Tschuggen Bergoase Wellness Center, Mario Botta Architetto, Arosa, Switzerland, 2006

The walkway made of aqua glass and white granite connects the five star vacation hotel built in the 1960s and the reception area on the third floor of the building of this century. A dark granite staircase will direct you to the third floor if you enter the building from the first floor. The different areas of "Berg Oase" are characterized by their interrelation and privileged relationship with the environment that guarantee natural lighting, an extraordinary sight towards the landscape and, at night, the building gains a magic atmosphere through artificial lighting.

材料：水蓝色玻璃、花岗岩

Materials: aqua glass, granite

亮蓝灰 Light Slate Gray

 闪兰色 Dodger Blue

 军兰色 Cadet Blue

© Urs Homberger

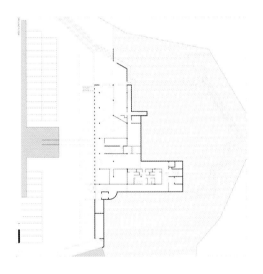
一层平面图 GROUND FLOOR PLAN

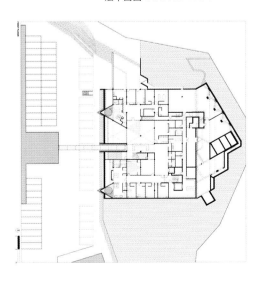
二层平面图 FIRST FLOOR PLAN

© Urs Homberger

希望之光
LIGHT OF HOPE

 纯蓝 Blue

 纯红 Red

 银灰色 Silver

 纯黑 Black

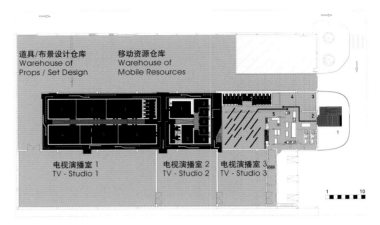

A座：
1 造型室
2, 4, 5, 6 更衣室
3 衣柜
7 卫生间
B座：
1 UPS中控室
2, 3 生产制作室
4 技术中心
5, 6 控制室
公共交流区域：
1 进入模块
2 接待区
3 等候室
4 量贩店
5 食堂/餐厅和厨房设施

Block A:
1 Make - Up / Hair
2, 4, 5, 6 Dressing Room
3 Wardrobe
7 Toilet
Block B:
1 UPS Central Room
2, 3 Production Room
4 Technical Central Room
5, 6 Control Room
Social Area:
1 Entry Module
2 Reception
3 Waiting Lounge
4 Merchandising Store
5 Cafeteria / Restaurant and Kitchen Facilities

一层平面图 GROUND FLOOR PLAN

© Paulo Segadães

© Paulo Segadães

SIC物流网络办公室，Marco Rocha，葡萄牙，卡纳西迪，2010年

该项目的一大特色在于半透明部门穿插的空间设计，使得自然光线不时射入室内。因此，建筑的中央区域是其建筑理念的根基。一个黑盒子般的体块依靠天窗射入的光线照明，并向两侧延伸出两个半月形的体量。整个项目实现了空间与空间、色彩与透视的协和共栖，使得建筑不仅是一个工作场所，而且可面向社会，供游人参观。配色方案与建筑本身及环境相得益彰。黑色的色调与天窗照明和人工照明形成对比。光线透过天窗洒向地面，仿佛给黑色的空间带来了希望之光。通道运用不同的颜色划分区域，给空间增添了活泼的气氛。

Logistics Complex of SIC Network, Marco Rocha, Carnaxide, Portugal, 2010

The project is characterized by its spatial coverage with some translucent parts, allowing the occasional use of natural light. As such, the concept is developed from the central zone. A black box partially illuminated by zenithal light emerging from promptly located skylights, and from where two crescent-shaped architectural volumes are developed. The entire project has a symbiotic dichotomy between space and place and between color and scenography, being that the architectural program has the peculiarity of being a space not only to work but also to visit and socialize. The color works then as a method of self architectural reference and of the place, where black acts as a counterpoint to both the zenithal and the artificial light. The light shines onto the ground from the skylights, as if bringing hope to the space. Different colors were employed to divide the zones, which creates a active atmosphere to the space.

材料：钢架结构、玻璃

Materials: steel structure, glass

© Paulo Segadães

诡秘几何
MYSTERIOUS GEOMETRY

 闪兰色 Dodger Blue

 海军蓝 Navy

 浅海洋绿 Light Sea Green

 暗淡灰 Dim Gray

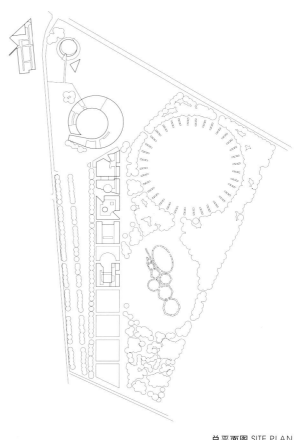

总平面图 SITE PLAN

© Thomas & Poul Pedersen

© Thomas & Poul Pedersen

卡尔·亨宁·彼得森 & 埃尔斯·阿尔菲尔特博物馆，C. F. Møller Architects，丹麦，海宁，2007年

博物馆最为过人之处便是以雕塑般的形体实现了艺术和建筑的统一。建筑的立面覆盖着瓷砖，上面绘有卡尔·亨宁·彼得森的彩色神兽。建筑整体呈几何形，圆形的主楼与对面改建后的工厂交相呼应。同时，整座建筑的几何形体也与文化和教育园区内的其他建筑相得益彰。

Carl-Henning Pedersen & Else Alfelt Museum, C. F. Møller Architects, Herning, Denmark, 2007

The museum is characterized by the unity of art and architecture in the sculptural form. The façade is clad with ceramic tiles decorated with Carl-Henning Pedersen's colorful mythical beasts. The shape is entirely geometric, with a circular main building echoing the shape of the converted factory situated opposite. The geometry of the entire complex also interacts with the other buildings in the cultural and educational park.

材料：瓷砖

Material: ceramic tiles

蓝色向往
THE BLUE DREAM

 纯蓝 Blue

 闪兰色 Dodger Blue

 银灰色 Silver

 秘鲁色 Peru

© Peter Mauss/ESTO

© Peter Mauss/ESTO

蓝色住宅，Bernard Tschumi Architects，美国，纽约，2007年

蓝色住宅是纽约城一座高约55.2 m的17层建筑。深具像素化效果的立面不仅反映出了其内部的空间布局，而且还代表了周围建筑的多元化特征。像素化的立面由约4 000块玻璃构成。其中包括构成幕墙的64块蓝色透明玻璃和2 400块不透明的蓝色拱肩玻璃。立面上使用了4种色度的蓝色绝缘拱肩玻璃。斜向的玻璃墙是建筑内许多公寓的一大特色。所有公寓的客厅和餐厅均装有落地玻璃窗。建筑标志性的像素拼镶立面映照着附近的各个社区，并与天空融于一色，与纽约市中心的喧嚣与动感相得益彰。

Blue Residential Tower, Bernard Tschumi Architects, New York, USA, 2007

Blue Residential Tower is a 55.2 m-high, 17-story structure built in New York. The pixelated façades reflect both the internal arrangement of spaces and the multi-faceted character of the neighborhood below. The pixelated façade consists of about 4,000 individual glass pieces. There are 64 blue-tinted vision glass pieces of the curtain wall and 2,400 opaque blue spandrel glass pieces. Four different shades of insulated blue spandrel glass were used to complete the façade design. The sloped window wall is a feature of many units. All units have full-height windows in the living and dining rooms. The building's signature pixelated mosaic façade reflects the various communities in the vicinity as well as one with the sky, in a way that defers both to echo and vibration dynamics of downtown New York City.

材料：玻璃

Material: glass

太阳能生态住宅
SOLAR RESIDENCE

 纯蓝 Blue

 米色 Beige

 栗色 Maroon

雪白色 Snow

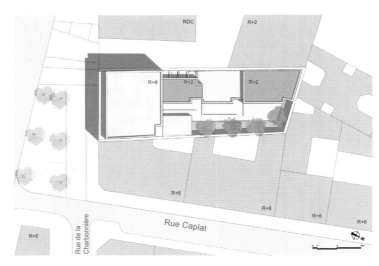

总平面图 SITE PLAN

© G. Kalt

巴黎18区社会住宅，Philippon-Kalt Architectes，法国，巴黎，2010年

建筑外墙覆有网格状的玻璃，并安装有太阳能电池板。建筑呈现双层表皮，蓝色太阳能电池板不对称地安置在玻璃外侧，能够保证隐秘性，同时能够阻挡阳光。这些面板吸收太阳光产生能量，提供居民热水所需能量的44%。涂色面板不仅能够阻挡视线进入公寓楼，同时也可以过滤并降低噪声。

Social Housing in the 18th District of Paris, Philippon-Kalt Architectes, Pairs, France, 2010

The project features a grid-like glass façade that has been fitted with energy-generating solar panels. Double-skin, tinted solar panels are installed in an asymmetrical pattern, with a dual function of providing privacy and shade. The panels soak up the sun's rays, generating 44% of the energy needed to heat water for the residents. And the tinted panels not only obscure views into the apartments, but also filter and reduce noise..

材料：玻璃、太阳能电池板

Materials: glass, solar panels

奢华盛宴
LUXURY FEAST

 纯蓝 Blue

 淡灰色 Gainsboro

 暗淡灰 Dim Gray

 暗灰蓝色 Dark Slate Blue

© Lin Ho

© Lin Ho

升禧艺廊购物中心，Spark Architects，马来西亚，吉隆坡，2011年

　　升禧艺廊堪称是吉隆坡最具代表性的购物商场，开设有一系列豪华的奢侈品商店和高级餐厅。Spark事务所对商场的原立面进行改造，设计着力打造开放的建筑立面形象，连续的店面将原建筑包裹在玻璃和石板构成的水晶状外墙里，吸引着人们前往。新立面酷似古希腊和古罗马雕像的"湿衣褶皱织物"和商场内出售的制作精美的礼服。实体与通透性的对比将原建筑的沿街立面彻底改变，赋予建筑全新的现代经典属性，从而使其从周围快速建造的众多购物中心中脱颖而出。

Starhill Gallery, Spark Architects, Kuala Lumpur, Malaysia, 2011

Starhill Gallery is perhaps Kuala Lumpur's most iconic shopping mall, featuring an extraordinary array of luxury shops and fine dining restaurants. Spark's design dealt with the reinvention of the existing façade of Starhill Gallery facing Bukit Bintang. The design has opened up the façade which provides a lot of visual interest via a continuous shop front that wraps the existing building in a crystalline skin of glass and stone panels. The new façade resembles the "wet drapery" of the ancient statues of Greece and Rome, and the beautifully crafted gowns on sale inside Starhill Gallery. The fractured variation of solidity and transparency transforms the street façade of the existing building entirely, giving it a new contemporary classic identity that stands out amongst the quick-fix, ubiquitous shopping mall façades of many of Starhill Gallery's neighbors.

材料：玻璃、石板

Materials: glass, stone panels

■赭,赤土也。——《说文》■赭褐色,代表着怀旧、平衡、深情、包容、温暖。尤其适合以木格调出现的地板、墙壁和家具,也适合于现今流行的锈蚀立面。■从淡淡的茶色到浓重的咖啡色,都是万能保险的颜色,适合社交和餐饮的公共场所,可以让墙壁、顶棚和地板轻松地变得温暖美观。■怀旧复古的建筑充满历史岁月的痕迹,装满难以磨灭的记忆。■如果使用不当,大面积的赭色可能显得消极沉重。

■ Ocher is the color of terracotta. ■ Ocher stands for nostalgia, balance, deep love, tolerance and warmth. It is particularly suitable for floors, walls and furniture in wood, and also the popular rusty façades. ■ The ocher colors from light tawny to dark coffee are almighty, suitable for public places such as communication and dinning spaces. They can easily warm and beautify walls, ceilings and floors. ■ Pseudo-classic architecture are filled with traces of history and loaded with indelible memories. ■ If used improperly, large areas of ocher may bring a negative and heavy feeling.

赭之怀旧
OCHER
321

花园中轻摇的风铃
AEOLIAN BELLS GENTLY SWINGING IN THE GARDEN

 巧克力色 Chocolate

 茶色 Tan

 重褐色 Saddle Brown

 雪白色 Snow

 暗绿色 Dark Green

索莫茹社区中心，Salto AB，爱沙尼亚，西维鲁县，索莫茹，2010年

　　索莫茹社区中心位于北爱沙尼亚郊外的拉克维小镇上，临近高速公路。该建筑的设计旨在区别于已有的建筑环境，而与周围的自然环境融为一体，采用鲜艳的秸秆般木梁搭配黑白混凝土外墙使之得以实现。教区决定将该社区中心设计为集多功能于一体的建筑来满足人们的不同需求。索莫茹社区中心的室内室外相互融合，制造出一种类似公园一样的结构体系，可爱繁茂的花园弥补了周围植被稀疏的地貌。整个社区中心都在一个平面上，不同高度的房间造就了起伏的屋顶。

Sōmeru Community Center, Salto AB, Sōmeru, Lääne-Viru County, Estonia, 2010

Community center is located Sōmeru just outside the North-Estonian town of Rakvere, on an open field near the highway in Sōmeru parish. The building was designed to stand apart from the existing built environment, more relating to the field surrounding it. This has-beens achieved by using colorful straw-like wooden beams attached to the black-and-white concrete façade. Instead of erecting buildings for their many different needs, the parish decided to combine multiple functions in one single building. Sōmeru Community Center combined parish administration, library and a hall with a club. The interior and exterior are intermingled, creating a park-like structure - cozy gardens as concentrated parts of nature offer a counterpart to sparse density of the surrounding environment. The community center is single-storeyed, undulating roofline is a result of varying heights of the rooms.

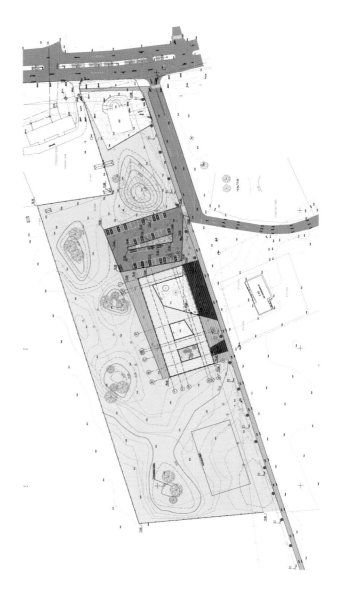

总平面图 SITE PLAN

材料：木材

Material: wood

© Courtesy of Salto AB

鲑鱼色 Salmon

橄榄褐色 Olive Drab

硬木色 Burly Wood

亮蓝灰 Light Slate Gray

东立面图 EAST ELEVATION　　西立面图 WEST ELEVATION

北立面图 NORTH ELEVATION

南立面图 SOUTH ELEVATION

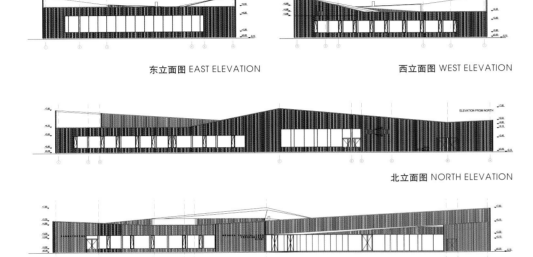

© Courtesy of Salto AB

索莫茹社区中心，Salto AB，爱沙尼亚，西维鲁县，索莫茹，2010年

类似的美学标准也应用到了室内设计，唯一不同的是室内的一根根木条是随意地悬挂在黑色顶棚上，似无数风铃，营造出一种活泼的氛围。橄榄色的窗帘、座椅，与室外的绿色草坪交相呼应，为室内增添温暖和生机。该单层建筑的构造安排得很紧凑，有足够大的公共活动区域和3个各具特色的开放庭院。

Sōmeru Community Centre, Salto AB, Sōmeru, Lääne-Viru County, Estonia, 2010

The same aesthetics continuous in the interior, only this time the slats are hanging freely from the ceiling, creating a lively accent above the black walls. The olive curtains and chairs respond to the green lawn outside, bringing much warmth and vitality to the interior. The single-storied building's many functions are arranged compactly, with enough room for common space and three open courtyards with different characters.

材料：木材

Material: wood

© Courtesy of Salto AB

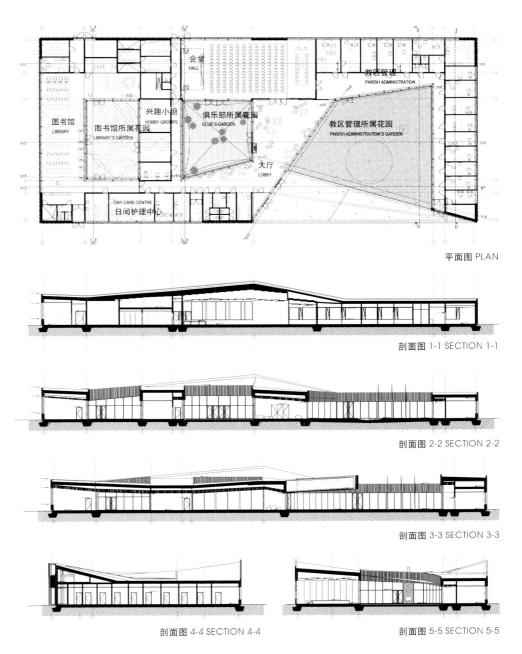

平面图 PLAN

剖面图 1-1 SECTION 1-1

剖面图 2-2 SECTION 2-2

剖面图 3-3 SECTION 3-3

剖面图 4-4 SECTION 4-4   剖面图 5-5 SECTION 5-5

锈蚀的回忆
RUSTY MEMORIES

 巧克力色 Chocolate

 银灰色 Silver

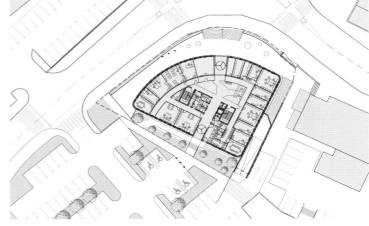

总平面图 SITE PLAN

耶瑟奈斯行政中心，Andrej Kalamar + Studio Kalamar，斯洛文尼亚，耶瑟奈斯，2012年

建筑的表皮与耶瑟奈斯当地钢铁制造业的传统形成呼应。即使在今天，城里的许多房子也都呈红锈色。这座行政中心的立面采用考顿钢，也呈现出锈蚀的特征。开口有节奏地镶嵌在建筑立面上，它们横向排列开来，形成了无限动感，与交通的动态感遥相呼应。建筑师采用考顿钢板、窗口系统、隔热系统等，构建出一个节能、低成本的建筑。

Administrative Center Jesenice, Andrej Kalamar + Studio Kalamar, Jesenice, Slovenia, 2012

The outer skin of the building establishes a dialogue with the steel-manufacturing industrial tradition of Jesenice. Even today, many of the houses in town are red-rusty, and the façade of the administrative center is rusty indeed due to its Corten skin. A pulsing rhythm of façade openings creates a horizontal dynamics, establishing a dialogue with dynamics of the traffic. By using efficient Corten panel shades, use of quality window systems and good thermal insulation the architects created an inert energy system which will ensure very low running costs.

北立面图 NORTH ELEVATION

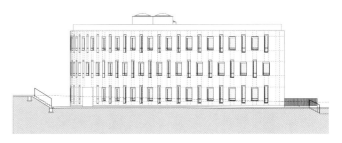

西立面图 WEST ELEVATION

材料：考顿钢

Material: Corten steel

东立面图 EAST ELEVATION

© Miran Kambič

南立面图 SOUTH ELEVATION

© Miran Kambič

秘鲁色 Peru

 雪白色 Snow

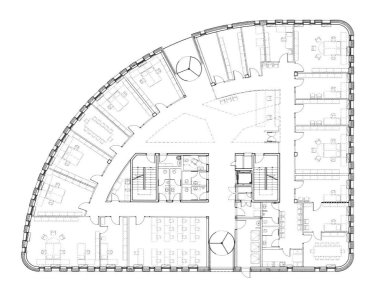

一层平面图 GROUND FLOOR PLAN

耶瑟奈斯行政中心，Andrej Kalamar + Studio Kalamar，斯洛文尼亚，耶瑟奈斯，2012年

受环境与光线的影响，白色的墙面呈现出淡淡的亮蓝色，使整个空间洁净明亮。层层木板环绕向上，宛若蜿蜒曲折的山间小道，吸引你前去探寻一番。自然光线、自然通风以及随意转换的空间都为员工创造了最佳的工作环境。所有的办公室均围绕室内的中庭排布。中庭自然采光充沛，并有天桥连接着各个空间。其美观的造型和温和的采光条件摒除了常规办公室沉闷的特征。立面系统将强烈的自然光线隔滤成更为柔和的漫射光线，促进了使用者全天采取自然光。建筑既与外部环境开敞连通，同时又默默地排除了影响使用者的所有因素。

Administrative Center Jesenice, Andrej Kalamar + Studio Kalamar, Jesenice, Slovenia, 2012

Influenced by the environment and light, the white wall is tinted with a light shade of blue, thus creating a clean, bright space. Layers of wooden slats circle upwards like the meandering hill roads which attract you to explore its detail. Natural light, natural ventilation, spontaneous transition of spaces are elements that create optimal working conditions for employees as well. All offices are organized around an interior atrium with connecting bridges and natural illumination, the attractive shape and friendly atrium illumination works to remove a stigma too often associated with state offices. Façade system filters the aggressive contrast light into a softer diffuse light, thus encouraging use of natural light throughout the day. Building remains open into the environment while at the same time unobtrusively filtering everything that affects the wellbeing of its users.

材料：木材、混凝土

Materials: wood, concrete

© Miran Kambič

© Miran Kambič

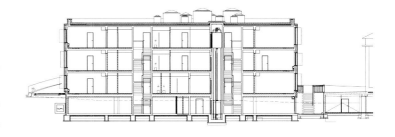

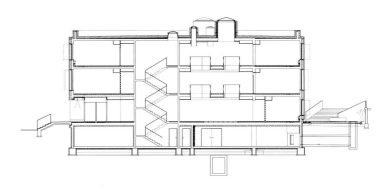

剖面图 SECTIONS

缤纷春季乐园
COLORFUL SPRING FAIRYLAND

 秘鲁色 Peru

 纯黄 Yellow

 暗黄褐色 Dark Khaki

 灰色 Gray

潘普洛纳幼儿园，Javier Larraz，西班牙，纳瓦拉，2009年

孩子在幼儿时期已经拥有巨大潜力，因此建筑师为其设计学校时需要深刻考虑责任，建筑师认为建筑设计应具有教学价值，能够形成合适的空间，使孩子们在接受教育的同时，能够以一种激励性的、有兴趣的以及安全的方式成长和发展。

设计师在空间中使用了不同的颜色和材料（混凝土，橡胶，圆形石头，草坪和木材）为孩子们创造了许多具有暗示性的多样的娱乐空间。建筑的西面装有网格结构。它由钢制的菱形板条组成，保护了建筑不受太阳光的直射。以黄色为基调的明快色彩，给人春天般的感觉。

Nursery School in Pamplona, Javier Larraz, Navarra, Spain, 2009

The impressive potential acquired by a child in its first years of life leads the architects to think deeply about the responsibility of designing an elementary school. The architects firmly believe in the pedagogical value of architecture and in its capability to generate spaces where, in combination with a proper educational task children can both grow and develop themselves in a stimulating, appealing and safe way during their first three years.

The use of different colors and textures (concrete, rubber, rolling stones, grass and wooded area) creates suggestive and varied playing spaces for children. West façade is protected against the direct evening sun through a structural lattice composed by vertical slats which are basically created by steel rhombus-shaped section pillars. The bright colors in yellow colors make people feel a spring-like warmth.

材料：混凝土、钢材

Materials: concrete, steel

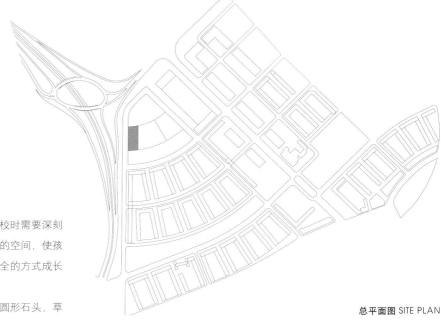

总平面图 SITE PLAN

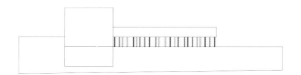

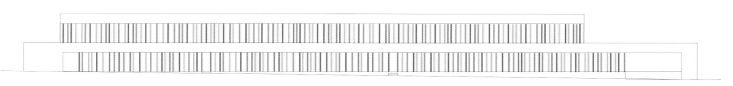

立面图 ELEVATIONS

© Iñaki Bergera

© Iñaki Bergera

雪白色 Snow

 绿黄色 Green Yellow

 秘鲁色 Peru

桃肉色 Peach Puff

潘普洛纳幼儿园，Larraz Arquitectos，西班牙，纳瓦拉，2009年

教室室内空间的组织以及家具的设计都考虑到了孩子们和教育者的不同感官能力。一方面，孩子们的活动区域根据不同的主题来划分，空间的设置非常灵活，他们可以在这里进行不同的活动；另一方面，教育者必须时刻能看到孩子们的行为和活动。

Nursery School in Pamplona, Larraz Arquitectos, Navarra, Spain, 2009

The organization of the inner space of the classrooms as well as the design of the furniture has been planned taking into account both children's and their educators' different perceptions of spaces. On one hand, children's activities are organized based upon several series of thematic "corners" perfectly adapted to their scale where they can develop different activities in a flexible way. On the other hand, educators must have children under visual control from any point of the classroom.

© Iñaki Bergera

© Iñaki Bergera

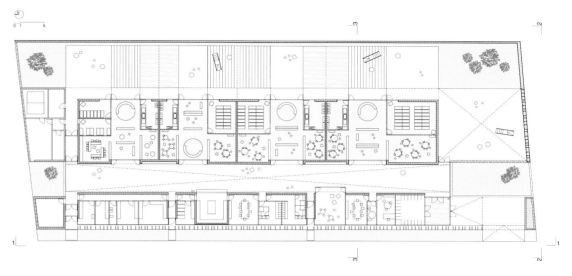

平面图 PLAN

© Iñaki Bergera

赭之怀旧 | OCHER

岁月的沉淀
SEDIMENT OF TIME

 黄土赭色 Sienna

纯黄 Yellow

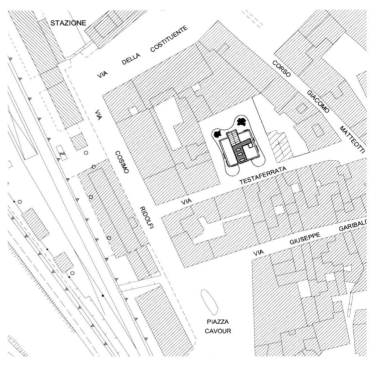

总平面图 SITE PLAN

© Alessandro Ciampi

贝诺佐·戈佐利博物馆，Massimo Mairani，意大利，佛罗伦萨，菲奥伦蒂诺堡，2009年

在当地政府的要求下，整座建筑都被红砖所覆盖，从而在材质和外饰上能够与当地的一些教堂相匹配。建筑严格遵沿已拆除的原建筑的平面，呈现了正方形般的外轮廓。建筑的基底形如岛屿一般，与地面贴合在一起，流线型的基底环托着整座建筑。室内存放着壁画 *Tabernacolo della Madonna della Tosse*。建筑规模较小但内秀其中，就像一个家庭工作室一样，似乎贝诺佐·戈佐利仍在埋头绘制我们所参观的壁画。

Benozzo Gozzoli Museum, Massimo Mairani, Castelfiorentino, Florence, Italy, 2009

Requested by the local municipality, the building is entirely coated in red bricks as it matches the materials and finishes of some local churches. The new design of the building closely follows the ground footprint of the demolished building, forming a square like shape. The building is rooted to the ground with a functional base, a shaped island. The curvilinear base runs around the building. *"Tabernacolo della Madonna della Tosse"* is displayed in the building which, due to its small size, retains something domestic. It's like a home studio where Benozzo Gozzoli seems to join us to visit these frescos while he's still working on them.

© Alessandro Ciampi

材料：红砖

Material: red bricks

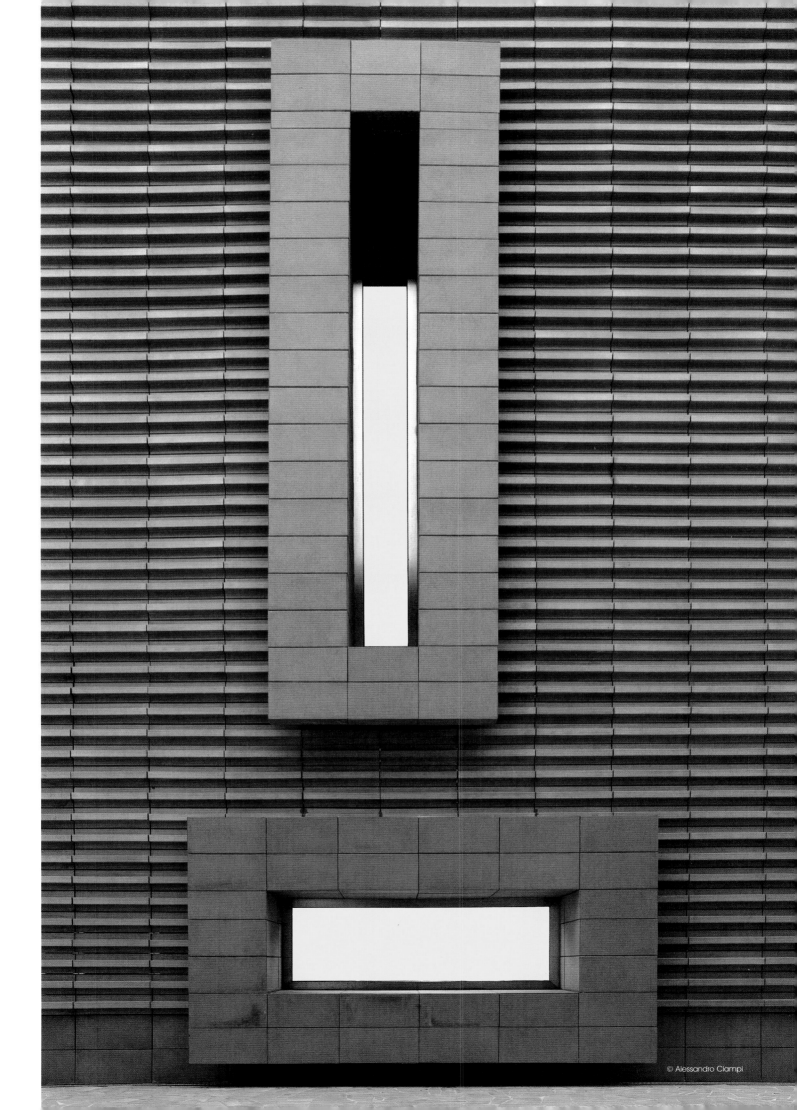

穿越古老的记忆
TRAVELING THROUGH THE OLD MEMORIES

 灰色 Gray

 暗黄褐色 Dark Khaki

 秘鲁色 Peru

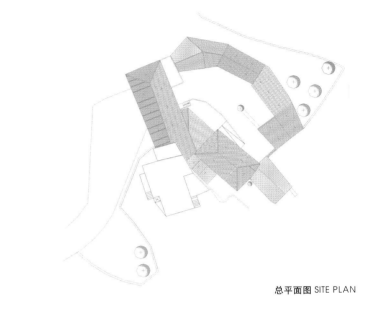

总平面图 SITE PLAN

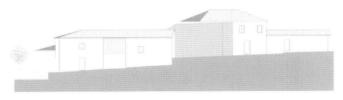

立面图 FAÇADE

森拉住宅，Manuel Ribeiro，葡萄牙，法菲，特拉瓦索斯，2010年

　　森拉住宅嵌立在米尼奥独具特色的自然环境之中，是葡萄牙古朴的乡村住宅遗产的生动范例。这座罕有的住宅仅有三种材料构成：木材、石头和陶瓷。天然的纹理与质朴的颜色，使住宅展现出一片宁静祥和的景色。建筑的主要功能区（住宅）浮现在山谷之中，刻意地与太阳的运动轨迹排成一线，保证了建筑的采光。建筑师称："我有意保持建筑原有的特性，决定沿用原始的窗户样式，甚至卫生间的窗户也是如此，以便在室内外产生丰富的光影效果。"

Senra´s House, Manuel Ribeiro, Travassós, Fafe, Portugal, 2010

Inserted in the distinctive nature of Minho, Senra´s House is a living example of the rustic countryside heritage in Portugal. The actual unfamiliar residence – exclusively made by three materials: wood, stone and ceramic. The natural textures and simple colors present a tranquil and peaceful image. The construction of the main functional body (housing), emerges in a valley, deliberately aligned with the Sun, ensuring that the building benefits from sunlight. "Having the intention to preserve the formal identity, it was decided to keep the windows in its original model, even in the toilets in order to play in various forms with the light of the interior and exterior …" justifies Manuel Ribeiro.

材料：木材、石头、陶瓷

Materials: wood, stone, ceramic

© Ivo Tavares

	灰菊黄 Pale Goldenrod
	秘鲁色 Peru
	暗淡灰 Dim Gray
	军兰色 Cadet Blue

平面图 PLAN

森拉住宅,Manuel Ribeiro,葡萄牙,法菲,特拉瓦索斯,2010年

通过对旧有结构的改造和新的木桁架的使用,所有的内部空间均呈现出新的面貌。古朴的石墙在室内也继续沿用,木门、屋顶、木质地板……仿佛童话故事中矮人的小屋般,展现在我们的眼前。此外,建筑的可持续性也不容忽视。这座新建筑有效利用了自然资源和再生能源,如:高效节能的LED灯、复合能源采暖炉、太阳能壁炉和火炉等。

Senra´s House, Manuel Ribeiro, Travassós, Fafe, Portugal, 2010

All the interior spaces reveal new features through the adaptation of old structures and deployment of a new system of coverage in wooden trusses. The pristine stone wall continues to appear indoors. Wooden door, roof, flooring…all are reminiscent of the dwarfs' house in the fairy tale. It should be noted, moreover, that on the level of sustainability, the new building has efficient use of natural resources and renewable energy, such as eco-efficient LED lights, heating system boiler mixture of pellets; solar panel fireplace and stove.

材料:木材、石头、陶瓷

Materials: wood, stone, ceramics

© Ivo Tavares

© Ivo Tavares

蒲公英的世界
A WORLD OF DANDELION

 灰菊黄 Pale Goldenrod

 橙色 Orange

 洋李色 Plum

 橙红色 Orange Red

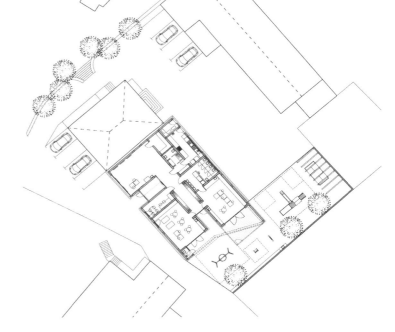

总平面图 SITE PLAN

立面图 ELEVATIONS

© Matevz Paternoster

塞露幼儿园，Minimart，斯洛文尼亚，塞露，2010年

  幼儿园采用柔和的颜色、彩绘和适当的比例，与周围环境融为一体。屋顶的大屋檐成为一处有趣的过渡空间，连通着室内外的空间。外立面采用纯度较低较温和的颜色，整体给人以古朴温暖之感，同时利用色彩条的渐变，增强了建筑本身的趣味性和可爱之处。外立面上绘制的表盘般的蒲公英，是花卉中独一无二的，就像幼儿园中的孩子们，它们簇拥在一起成长，待到成熟了，就各自拥有了自己的"小伞"，四散开来，自由自在地飞向广阔的新世界。

Kindergarten Selo, Minimart, Selo, Slovenia, 2010

 The building blends seamlessly with the surrounding structures due to its pastel façade and its proportions. The shifted roof of the atrium provides for an interesting transition between interior and exterior. The pastel hues of the façade are unobtrusive and subtle, thus creating a pristine and warm atmosphere. Meanwhile, the usage of a color ramp further highlights the fun and cuteness of the building itself. The dandelion clocks on the façade are a symbolic decoration, representing that most unique among flowers, unique as the children within the kindergarten, and the growth from flower to blowball symbolizes those first steps of freedom that usher the way into a wider world.

材料：混凝土、涂料

Materials: concrete, paint

© Matevz Paternoster

浅鲑鱼肉色 Light Salmon

黄土赭色 Sienna

深灰色 Dark Gray

暗淡灰 Dim Gray

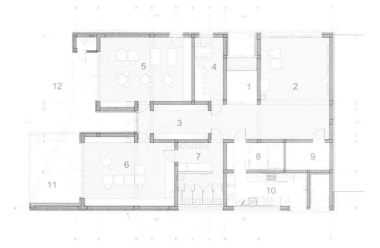

1 入口　　　　1 Entrance
2 多功能厅　　2 Multipurpose Room
3 更衣室　　　3 Wardrobe
4 储物间　　　4 Cabinet
5 儿童游戏室2　5 Children's Playroom 2
6 儿童游戏室1　6 Children's Playroom 1
7 卫生间　　　7 Toilets
8 员工室　　　8 Staff
9 商店　　　　9 Store
10 厨房　　　10 Kitchen
11 露台　　　11 Terrace
12 操场　　　12 Playground

平面图 PLAN

© Matevz Paternoster

塞露幼儿园，Minimart，斯洛文尼亚，塞露，2010年

室内的墙壁上则绘制了一些粉彩的装饰性图画。空间上延续了外立面设计的趣味性，墙面铺设的瓷砖也设计为彩条的装饰性图案，成功地保留了设计师丰富的想象力和创新精神，为孩子们创造出了一个流连忘返的温馨环境。

Kindergarten Selo, Minimart, Selo, Slovenia, 2010

The interior walls also have several decorative prints on them. The interior space is a continuation of the interesting façade design. The ceramic tiles, forming several colorful patterns, succeeded in retaining the luxuriant imagination and innovative spirit of the architects. It creates an environment to which children gladly return.

材料：瓷砖、涂料

Materials: ceramic tiles, paint

© Matevz Paternoster

返璞归真
RETURNING TO THE ORIGINAL NATURE

 金菊黄 Goldenrod

 重褐色 Saddle Brown

 橙红色 Orange Red

 纯红 Red

 中海洋绿 Medium Sea Green

 纯黄 Yellow

 中蓝色 Medium Blue

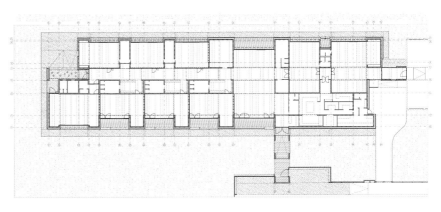

平面图 PLAN

佛萨伦噶幼儿园，Studiomas Architetti Associati，意大利，特雷维索，2009年

幼儿园坐落在小镇的北部，位于楼房与自然风景之间，为这片区域勾画了新的天际线。红砖与木条构筑的外立面，颜色协调，给人古朴自然之感。不同颜色的色块点缀其中，增添了几分清新的味道。整座建筑与周围的环境融为一体，似一缕春风，轻抚大地。学校围绕中央大厅分布，南面朝向小镇，北面近临大自然。与大厅相连的小路，一端连着教室，另一端通向饭厅，一直延伸到老校区。这些小路的尺寸、形状、颜色、功能都各不相同，使其更容易区分。

Nursery School in Fossalunga, Studiomas Architetti Associati, Treviso, Italy, 2009

The nursery school is situated in the northern part of the village between the buildings and the landscape; the school itself draws the new horizon. The façade composed of red bricks and wooden slats is harmonious in color, presenting a pristine and natural style. Dotted with different color blocks, the façade gains a fresh favor. The building integrates with the surrounding, which as natural as the spring breeze caressing the earth. The school is designed around a central hall, that looks towards the village (South) and the landscape (North). The streets (passages), crossing the hall, connect, on one side, to the classrooms and, on the other side, to the dining-room till the old existing school. These passages are different in terms of dimension, proportion, color, function, which makes them more distinguishing.

© Marco Covi

材料：木材、红砖

Materials: wood, red bricks

© Marco Covi

剖面图 SECTIONS

成熟的果实
RIPE FRUIT

 亮菊黄 Light Goldenrod Yellow

 橙红色 Orange Red

 亮蓝 Light Blue

总平面图 SITE PLAN

东南立面图 SOUTHEAST ELEVATION

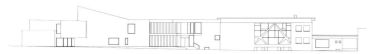

西北立面图 NORTHWEST ELEVATION

东北立面图 NORTHEAST ELEVATION

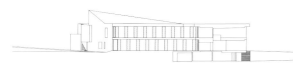

西南立面图 SOUTHWEST ELEVATION

沃戈斯克利中学扩建，Studio Granda，冰岛，2005年

由于地势较低，楼梯的外部镶满了全高式玻璃窗，其间穿插着细长的隔栅窗。隔栅窗采用了3种不同的颜色，在白色墙面的映衬下显得尤为突出，富有韵律感，使建筑的整体色调变得轻柔而温暖。但北立面是一个例外，由于它面向嘈杂的Skeidavogs大街，因而采用了更为厚实和美观的混凝土材质。大厅的天窗为建筑的中心部分供给了更多的太阳光。

Vogaskoli Secondary School Extension, Studio Granda, Iceland, 2005

As a consequence of the deep plan the periphery is predominately glazed with full height windows interspersed with attenuated grilles. The attenuated grilles are in three colors. They are even prominent against the white wall and create a rhythm, due to which the tone of the building becomes soft and warm. The exception is the more massive, fair-faced concrete, north façade due to the proximity of the noisy Skeidavogs road. The heart of the building receives additional daylight through the clearstory windows in the hall.

© Studio Granda

材料：玻璃、混凝土

Materials: glass, concrete

© Studio Granda

橙红色 Orange Red

 秘鲁色 Peru

 暗淡灰 Dim Gray

© Studio Granda

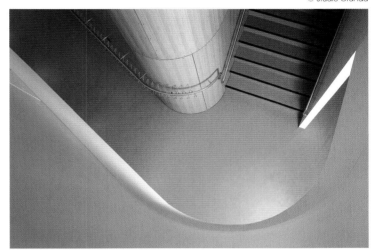
© Studio Granda

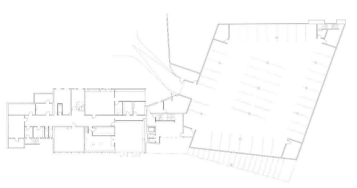
一层平面图 GROUND FLOOR PLAN

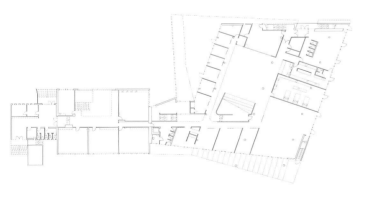
二层平面图 FIRST FLOOR PLAN

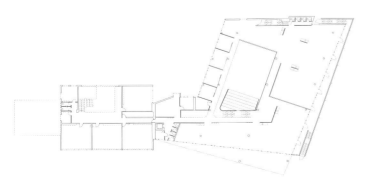
三层平面图 SECOND FLOOR PLAN

沃戈斯克利中学扩建，Studio Granda，冰岛，2005年

建筑内部的色调也相对统一，楼梯处运用木材本身的颜色搭配暖洋洋的橙红色，室内隔断设计成黑色，使环境沉稳、肃静，能瞬间让人平静。为了最大限度地发挥结构的灵活性，建筑采用了气泡板板材所构筑的双向中空平台和最少量的圆柱。屋顶采用了同样的结构，并覆盖了一层草皮。

Vogaskoli Secondary School Extension, Studio Granda, Iceland, 2005

The colors of the interior are relatively uniform. Orange red and the natural color of the wood are applied to the stairs while black to the partitions. A steady, solemn and silent environment is created which can calm people down in no time. The structure is of two-way hollow deck "bubbledeck" plates with a minimum of columns to maximize flexibility. The roof uses the same structure and is finished with a layer of turf.

材料：木材、气泡板板材

Materials: wood, "bubbledeck" plates

沁人心脾的温暖
THE SUNSHINE WARMTH BRINGS JOY TO THE SOUL

 象牙色 Ivory

 浅灰色 Light Gray

 金色 Gold

奥胡斯工程学院，C. F. Møller Architects，丹麦，奥胡斯，2011年

　　金黄色的玻璃板在蔚蓝色天空与灰白色墙面的映衬下，显得尤为突出，夺人眼球。阳光透过玻璃板，又把黄色映衬到墙面上，产生出富有韵律感的折印。纤长的窗框、金黄的颜色与墙面上的半透明折印相互交叠……使整座建筑都洋溢出高贵的气质。

The New IHA Engineering College of Aarhus, C. F. Møller Architects, Aarhus, Denmark, 2011

Set off by the blue sky and gray walls, the golden glass panels appear pretty distinctive and eye-catching. Sunshine penetrates the glass panels and tints the walls with golden color, creating rhythmic creases. In the interior, the slender window frames, the golden color and the translucent creases on the walls overlap each other, which bring nobility to the entire building.

材料：玻璃、混凝土

Materials: glass, concrete

© Helene Høyer Mikkelsen

© Helene Høyer Mikkelsen

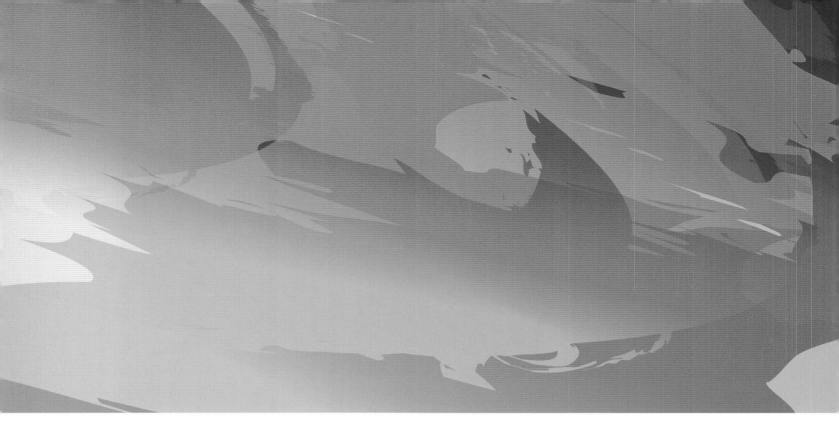

■赤者，火色也。——《洪范五行传》■红色，激情四溢的色彩，气势盛大，是热烈的象征。暗示能力、热情、生命、喜庆、乐观，亦代表进攻和反叛。可刺激身体直接反应，有助于体育锻炼或者激发积极性。■高纯度的红色，通常作为一种突出色少量使用，大面积使用时，由于能量巨大而气场过于强大。但其变化色，比如玫瑰色，栗色，珊瑚色和粉红色却被广泛地应用于彰显女性化、温柔的空间中，大面积的使用亦很出彩。

■ Red is the color of fire. ■ Red contains intense motion and great power, and it is an embodiment of passion. It represents power, passion, life, festivity, optimism, as well as aggression and rebellion. Red color can stimulate a person's body, help to promote exercises and arouse enthusiasm. ■ The red color of high purity is often used as an embellishment color to highlight the theme; if used in large scale, it will give a strong power since it contains too much energy. However, red has many variations, such as roseous, sorrel, coral and pink. They are wildly used in the feminine and tender space, and it looks good for the large area.

赤之热烈
RED
353

逐光追影
RED COLOR DANCE WITH LIGHT AND SHADOW

 浅灰色 Light Gray

 橙红色 Orange Red

 闪光绿 Lime

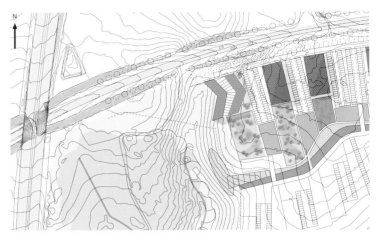

总平面图 SITE PLAN

© Julian Weyer

© Julian Weyer

咨询楼，C. F. Møller Architects，丹麦，瓦埃勒，里舒特商务园，2009年

当人们从高速公路疾驰而过，建筑非同寻常的几何形体会呈现出悦目而多变的外观效果。立面的色彩和纹理的设计基于对光线的捕捉，更加突显了建筑形体及光影变化。13种不同比例的外墙板"随意地"构成了立面上的条带，有些条带对角折拢成三角形。这些板材横向间隔错列，形成了色彩与光影的闪亮阵列。这些外墙板由铝材制成，并覆有特殊的色彩涂层。它们的色彩不断变幻，会随着视角和太阳角度的变化而呈现高光和有趣的色彩渐变。因此，建筑外观呈现多种变化。当从高速公路驶过时，这种效果尤为显著。

Advice House, C. F. Møller Architects, Lysholt Park, Vejle, Denmark, 2009

The building's unusual geometry makes for a dramatic and changing appearance when driving by on the motorway, and this mutability in form and shadows is further heightened by the coloring and texturing of the façades, designed to catch the light. The cladding-strips are composed of a "random" sequence of a total of 13 differently proportioned cladding panels, some of which are folded diagonally to create a triangulated pattern. The panels are mounted horizontally at staggered intervals, creating a glittering array of colors, light and shadows. The cladding panels are made from aluminum with a special color pigmentation that offers changing color effects with highlights and interesting color gradients, depending on the viewing angle and the angle of the sun. Thus, the building never appears in quite the same way, and the effect is especially striking when passing by on the motorway.

材料：铝板、玻璃

Materials: aluminum panels, glass

© Julian Weyer

西立面图 WEST ELEVATION　　　南立面图 SOUTH ELEVATION　　　东立面图 EAST ELEVATION

© Julian Weyer

© Julian Weyer

© Julian Weyer

© Julian Weyer

纯黄 Yellow

 橙色 Orange

 闪光绿 Lime

 硬木色 Burly Wood

 暗瓦灰色 Dark Slate Gray

© Julian Weyer

© Julian Weyer

© Julian Weyer

咨询楼，C. F. Møller Architects，丹麦，瓦埃勒，里舒特商务园，2009年

咨询楼内部是5 000 m²开放、灵活的办公布局，形成良好的视野和角度。内部以白色为主体，搭配红、黄、绿等明亮色彩，具有醒目的空间装饰效果，同时使人产生愉悦的视觉感受。

Advice House, C. F. Møller Architects, Lysholt Park, Vejle, Denmark, 2009

The Advice House interior is 5,000 m² of open and flexible office layout, which offers dramatic perspectives and angles. The interior is mainly white, complemented with bright colors such as red, yellow and green. These striking colors decorate the space, giving people a pleasant visual sensation.

材料：染色混凝土、竹地板、瓷砖地板、金属顶棚、吸声板

Materials: painted concrete, bamboo floors, ceramic floor tiles, metal ceilings, acoustic absorbent panels

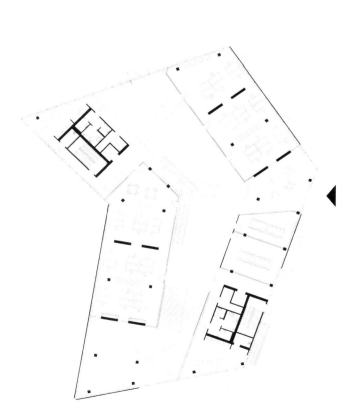

平面图 PLAN

热烈的韵律
ARDENT RHYTHMS

 金色 Gold

 橙色 Orange

 深橙色 Dark Orange

 橙红色 Orange Red

 深红色 Dark Red

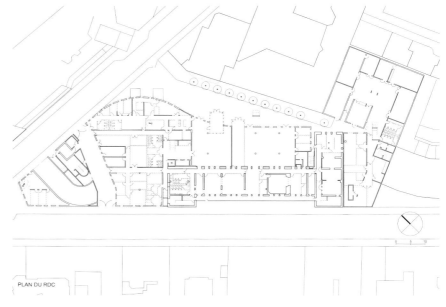

总平面图 SITE PLAN

© Takuji Shimmura

© Georges Fessy

诺凡西亚商学院，Architecture-Studio，法国，巴黎，2011年

整修后的建筑因其玻璃立面的优雅设计而别具一格。立面是由彩色印刷玻璃制成的活动垂直百叶组成的，呈现出红、黄渐变色调，韵律感十足。立面的颜色从红色渐变为黄色或从黄色渐变为红色，渐变的速度时快时慢。如此一来，7种颜色、双面的3种图案、7种不同的高度就提供了873种选择，最终形成4 102片百叶。颜色的渐变使建筑与周围的环境融合在一起，同时，由于玻璃立面映射出周围的景象，建筑本身也成为一种"环境"。这座现代、严谨、活泼及色彩斑斓的建筑从周围建筑中脱颖而出。建筑与周围的环境形成色彩的和谐，与周围的亮度和颜色形成互动。其精巧的设计更使得建筑活力无限。

Novancia Business School, Architecture-Studio, Paris, France, 2011

The renovated building is characterized by the refined design of glass façades composed of mobile vertical shutters with colored printed glass with a gradation from red to yellow, creating a strong sense of rhythms. Colors vary within a gradation going from red to yellow and from yellow to red. Gradations accelerate and decelerate. Seven colors, three patterns, on both sides, and seven different heights result in 873 references and 4,102 implemented items. Whereas color gradation makes the building contextual, it becomes the "context" itself due to the reflection of surrounding elements onto the façade. A contemporary, precise, joyful and colored building stands out of the street building alignment. The building is in chromatic harmony with nearby environment, and plays and interacts with the brightness and colors of its surroundings. Its refined design makes it even more dynamic.

材料：彩色印刷玻璃

Material: colored printed glass

诗意的栖居
POETIC DWELLING

 猩红 Crimson

 橙色 Orange

 闪光深绿 Lime Green

 深灰色 Dark Gray

埃皮奈学生公寓，Emmanuel Combarel Dominique Marrec architectes(ECDM)，法国，埃皮奈，2008年

公寓楼的设计理念是对当地道路景观的重新解读，将改造与历史传统相结合，以其外形和质量反映出场所具有的诗意。该项目形成"双面"效果，北面临街的一侧呈现立面的都市风格，而南侧则呈现不同的色彩，颇具乡村田园风格。

Student Housing in Epinay, Emmanuel Combarel Dominique Marrec architectes (ECDM) , Epinay, France, 2008

The conception of the project joins in a second reading of the landscape of the road of Saint leu, by integrating its history and its transformations to assert the manners and the qualities and reveal the poetry of the place. The project will be urban façade on the road side in the north, to become more paysagé and coil up in the continuity of fragmented one suburban very structured in the south.

材料：彩色印刷玻璃

Material: colored printed glass

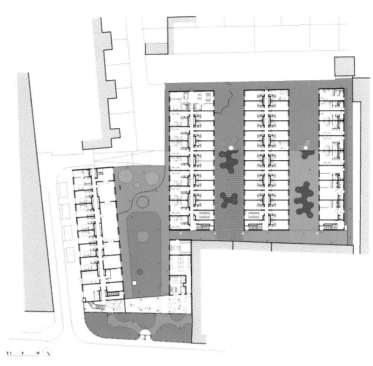

总平面图 SITE PLAN

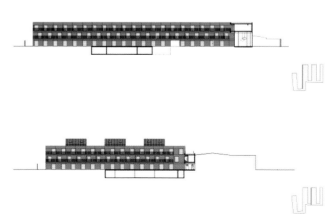

© Benoit Fougeirol

立面图 ELEVATIONS

© Benoit Fougeirol

© Benoit Fougeirol

美妙童梦
A MAGICAL DREAM FOR CHILDREN

 猩红 Crimson

 秘鲁色 Peru

 重褐色 Saddle Brown

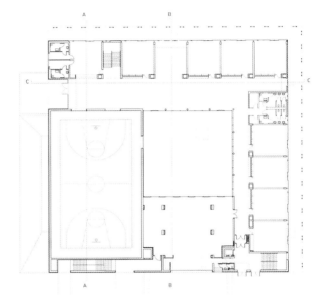

平面图 PLAN

© Alessandra Bello

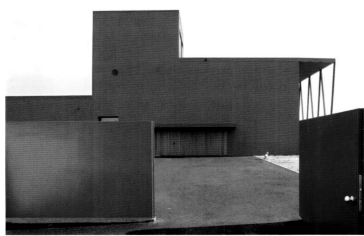
© Alessandra Bello

意大利庞萨诺小学,C+S Associati / Carlo Cappai, Maria Alessandra Segantini,意大利,庞萨诺威尼托,2009年

在该项目中,建筑师通过威尼托区红色的"谷仓"(紧邻由Campo Baeza设计的庞萨诺幼儿园,被解读为一种现代别墅)唤起人们的记忆。而建筑的当代特色则与贝纳通工厂和其设计文化及风靡全球的色彩民主哲学存在一定联系。从这层意义上说,这所学校则成为了孩子们心目中梦境或故事背景。故事的主人公是孩子和他们的老师,以及学校周围的人们。

Ponzano Primary School, C+S Associati / Carlo Cappai, Maria Alessandra Segantini, Ponzano Veneto, Italy, 2009

In this project memory is represented by a remind to cultivation, the red colored "barchessas" of the Veneto Region (near to the Campo Baeza Ponzano Children which is read as a kind of Contemporary Villa). The contemporary is linked to Benetton Factories and their culture of good design and philosophy spreading color democracy all over the world. In this sense the school architecture becomes the scenography of a dream (of a story would the children say) where the main characters are the children together with their teachers and the community around the school.

材料:金属、玻璃、木材

Materials: metal, glass, wood

© Pietro Savorelli

黄绿色 Yellow Green

 暗淡灰 Dim Gray

© Pietro Savorelli

意大利庞萨诺小学，C+S Associati / Carlo Cappai, Maria Alessandra Segantini，意大利，庞萨诺威尼托，2009年

在室内，色彩用来区分不同的空间，绿色的是流通空间，深灰色的是各种专用空间，天然木色及各种图案用来识别其他空间。

Ponzano Primary School, C+S Associati / Carlo Cappai, Maria Alessandra Segantini, Ponzano Veneto, Italy, 2009

Inside the color becomes a code to use the space: green is the color of relation spaces and dark gray that of special spaces. Natural wood and graphic design is used to recognize the other spaces.

材料：金属、玻璃、木材

Materials: metal, glass, wood

© Pietro Savorelli

透视剖面图 PERSPECTIVE SECTION

纳瓦白 Navajo White

硬木色 Burly Wood

深红色 Dark Red

© Alessandra Bello

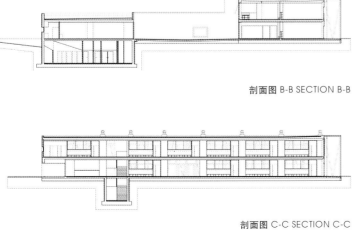

剖面图 A-A SECTION A-A

剖面图 B-B SECTION B-B

剖面图 C-C SECTION C-C

© C+S Associati

赤之热烈 | RED | 365

活力橙色空间
DYNAMIC ORANGE SPACE

 猩红 Crimson

 橙红色 Orange Red

 浅玫瑰色 Misty Rose

 黛青色 Dark Cyan

波士顿大学学生公寓的"断层"咖啡厅，Tony Owen Partners，澳大利亚，悉尼，2011年

　　咖啡厅的顶棚设计独特，其灵感源自断裂峡谷，是学生公寓楼的主要设计元素之一。这些"峡谷"造型最大限度地保证阳光照射和自然通风，同时，五光十色的玻璃确保了学生的隐私性。顶棚的几何形状取自一张公寓楼"峡谷"状窗户的照片，窗户采用橙色有机玻璃，形态各异。这张图片同时也作为图像呈现在咖啡厅的墙壁上。咖啡厅位于公寓楼旁边，整体造型活力无限。

Fractal Café in Boston University Student Housing, Tony Owen Partners, Sydney, Australia, 2011

The unique design of the ceilings is derived from the fractal canyons which are the main design element in the building. These canyons maximize solar access and natural ventilation, whilst the prismatic windows maintain privacy for the students. The geometry of the ceilings came from a photograph of the completed canyon windows. The ever changing geometry of these windows was interpreted here using orange plexi-glass. This photo was also used as a super-graphic on the café wall. The overall effect is a dynamic geometry which sits alongside the building itself.

材料：有机玻璃

Material: plexi-glass

© Courtesy of Tony Owen Partners

© Courtesy of Tony Owen Partners

© Courtesy of Tony Owen Partners

一米阳光
A RAY OF SUNSHINE

 猩红 Crimson

 橙红色 Orange Red

 橙色 Orange

象牙色 Ivory

东村公寓，Bercy Chen Studio LP，美国，得克萨斯州，奥斯汀市，2010年

建筑立面像是撒上一道阳光，呈现鲜艳的红色和橙色，在奥斯汀市东11街振兴区创造出强而有力的地标。建筑的设计创造性地采用节能材料。为了阻挡得克萨斯州强烈的阳光，南面和西面的外墙覆盖着多层面的钢屏，同时还可以兼做遮阳设备和阳台护栏。巴西艺术家Helio Oiticica设计的这些彩色屏面使该项目在城市景观中呈现标志性的视觉形象。

The East Village, Bercy Chen Studio LP, Austin, Texas, USA, 2010

Like a sunburst across the corner façades, the building creates a strong presence in Austin's East 11th Street Revitalization District with a bright red and orange façade. The design of the building incorporates modest materials in creative and dynamic ways. The upper south and west façades are shielded from the hot Texas sun with a multi-faceted steel panel screen which doubles as a shading device and guardrails for the residential balconies. Brazilian artist, Helio Oiticica, inspired the design of these colored screens, which gives the project its iconic visual presence in the urban landscape.

材料：钢板、玻璃

Materials: steel panels, glass

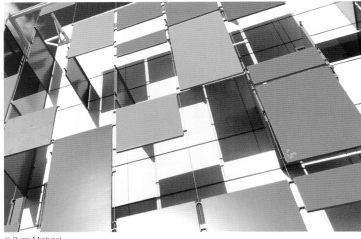
© Ryan Michael

© Ryan Michael

© Ryan Michael

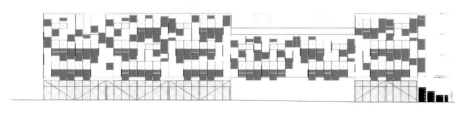
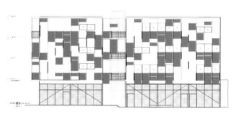

西立面图 WEST ELEVATION   南立面图 SOUTH ELEVATION

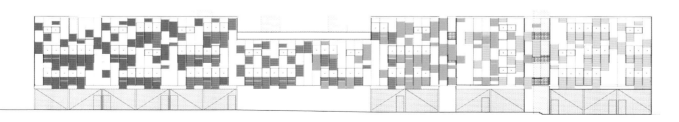

面板布局 PANEL LAYOUT

飘香热咖啡
THE WAVES OF HOT COFFEE

 纯红 Red

 银灰色 Silver

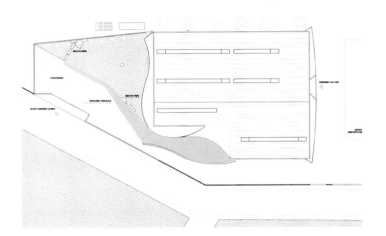

总平面图 SITE PLAN

© Angelo Margutti & Associati

© Angelo Margutti & Associati

立面图 ELEVATIONS

咖啡机博物馆，Arkispazio，意大利，米兰，2012年

咖啡机博物馆是为了庆祝金佰利集团100周年纪念而设计的，金佰利是世界最大的专业用咖啡机制造商。建筑外墙选用红色的带状装饰，象征热咖啡飘动的热气。夜晚，在背景光的照射下形成醒目的图案，使博物馆焕发活力。

Museum MUMAC, Arkispazio, Milan, Italy, 2012

The Museum MUMAC (Museum of Coffee Machine) is designed to celebrate the 100th anniversary of Cimbali Group, the most important professional coffee machine manufacturer in the world. The façades of the museum are covered with strips of metal "red Cimbali", sinuous and enveloping to resemble the waves of hot coffee, which at night filters the artificial light creating a striking illuminated reticle that evokes the energy of MUMAC.

材料：金属

Material: metal

© Angelo Margutti & Associati

万花筒般的世界
A KALEIDOSCOPIC WORLD

 纯红 Red

 橙红色 Orange Red

 橙色 Orange

 烟灰色 Smoke Gray

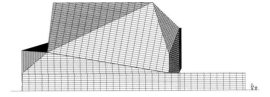

南立面图 SOUTH FAÇADE

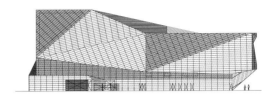

东立面图 EAST FAÇADE

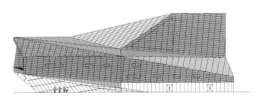

西立面图 WEST FAÇADE

北立面图 NORTH FAÇADE

Agora 剧院，UNStudio，荷兰，莱利斯塔德，2007年

剧院的所有立面均含有尖锐的角和突出的棱，由钢板和玻璃所覆盖，分层堆叠，呈黄色和橘红色。其设计关注剧院的基本功能——创造出一个巧妙、魔幻的世界。剧院的内外墙面都经过特殊设计，形成了一个万花筒般的世界。身处其中，你将分不清什么是真实的，什么是虚幻的。这里的戏剧和表演并不只限于舞台上，更延伸到都市生活中；不只限于夜晚，也延伸至白天。

Theater Agora, UNStudio, Lelystad, the Netherlands, 2007

All of the façades of the theater have sharp angles and jutting planes, which are covered by steel plates and glass, often layered, in shades of yellow and orange. The design focuses on the archetypal function of a theatre: that of creating a world of artifice and enchantment. Both inside and outside walls are faceted to reconstruct the kaleidoscopic experience of the world of the stage, where you can never be sure of what is real and what is not. In Theatre Agora drama and performance are not restricted to the stage and to the evening, but are extended to the urban experience and to daytime.

材料：钢板、皱纹铝板、铝网、玻璃

Materials: steel plates, corrugated aluminum, aluminum mesh, glass

© Christian Richters

© Christian Richters

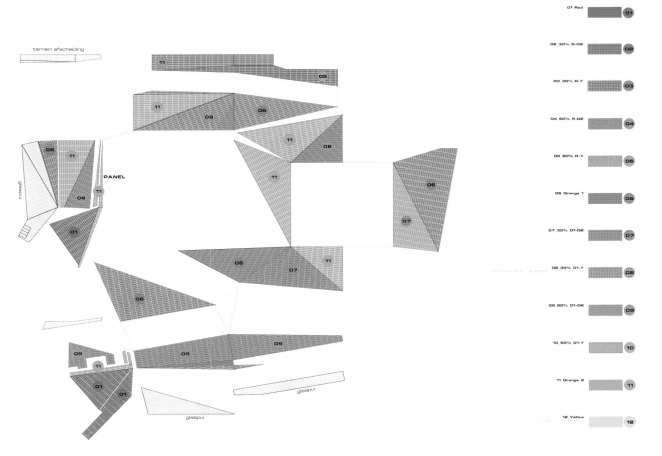

立面展开图 FAÇADE FOLDOUT

亮粉红 Light Pink

深粉红 Deep Pink

桃肉色 Peach Puff

© Iwan Baan

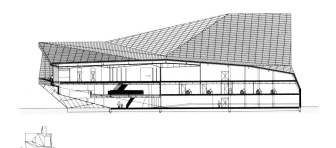

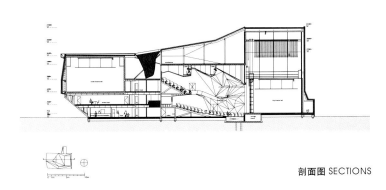

剖面图 SECTIONS

© Iwan Baan

Agora 剧院，UNStudio，荷兰，莱利斯塔德，2007年

外部的缤纷色彩在剧院内部得以延续和加强。栏杆就像是一条曲折的粉红缎带，沿着楼梯蜿蜒而下，围绕一层大型、开放门厅的中庭盘旋环绕，然后又沿墙壁蜿蜒而上直达屋顶。其颜色不断变化，由紫色变为深红再变为鲜红，最后几乎成了白色。

Theater Agora, UNStudio, Lelystad, the Netherlands, 2007

Inside, the colorfulness of the outside increases in intensity; a handrail executed as a snaking pink ribbon cascades down the main staircase, winds itself all around the void at the center of the large, open foyer space on the first floor and then extends up the wall towards the roof, optically changing colors all the while from violet, crimson and cherry to almost white.

© Iwan Baan

© Iwan Baan

纯红 Red

 深红色 Dark Red

© Iwan Baan

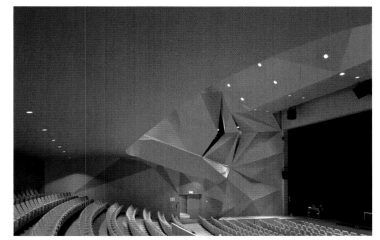
© Iwan Baan

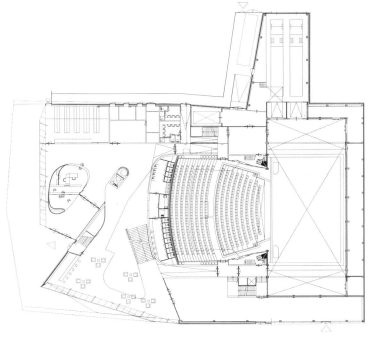
一层平面图 FIRST FLOOR PLAN

Agora 剧院，UNStudio，荷兰，莱利斯塔德，2007年

剧场主色调为红色，象征着艺术家们无法熄灭的艺术火焰。红色的内墙衬盖着呈凹凸状的隔声板，形状、颜色不同，不仅是为了追求视觉效果，同时也是为了改善剧场的声效。

Theater Agora, UNStudio, Lelystad, the Netherlands, 2007

The main theatre is all in red, representing the inspiration sparks of the artists. The red interior walls are lined with acoustic paneling of a concave / convex formation in various shapes and colors, which serves not only as a visual interest but was found by Dutch acoustics designers DMGR to also benefit the sound in the auditorium.

材料：吸声板

Material: acoustic panels

青春飞扬活力四射
VIGOR OF YOUTH

 深橙色 Dark Orange

 纯黄 Yellow

 暗橄榄绿 Dark Olive Green

雪白色 Snow

平面图 PLAN

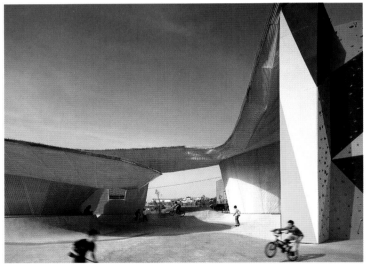
© Roland Halbe

© Iwan Baan

梅里达工厂青年活动中心，Selgascano，西班牙，梅里达，2011年

梅里达工厂青年活动中心是一个充满精彩的地方，就如同它呈现的颜色一样，动感亮丽、丰富多彩。这里的活动设施一应俱全，可以举行各式各样的活动。该建筑被设计成一个巨大的天篷，它向整座城市张开怀抱，可供任何人来此休闲。这个半透明的保护性屋顶就像一朵轻盈的云彩铺展开来。它由三维网格结构构成，厚1 m，包含多个层面。天篷被卵形部分边缘的钢柱所支撑，并在最高处与用同样的三维网格所构成的攀岩墙连接在一起。

Merida Factory Youth Movement, Selgascano, Merida, Spain, 2011

Merida Factory Youth Movement is as wonderful as the colors it presents - dynamic and colorful. In addition, it has well-equipped facilities for all kinds of activates. The building is conceived as a large canopy opened to the whole city to gather anyone who may need to shelter there. The roof is understood, and extends, like a light cloud, protective, translucent; constructed with a three-dimensional mesh structure 1 meter thick covering different levels. This canopy will rest on steel columns placed in the perimeter of the supporting ovoid elements up to the highest level, where it joins the Climbing Walls structure made with the same three-dimensional mesh.

材料：聚碳酸酯板、钢材

Materials: polycarbonate panels, steel

© Selgascano

© Selgascano

为生活空间增添活力
IMPROVING THE LIVING SPACE WITH VITALITY

 猩红 Crimson

 巧克力色 Chocolate

雪白色 Snow

 闪光深绿 Lime Green

© Frederik Vercruysse

© Frederik Vercruysse

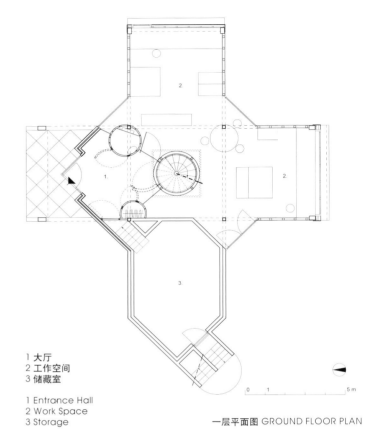

1 大厅
2 工作空间
3 储藏室

1 Entrance Hall
2 Work Space
3 Storage

一层平面图 GROUND FLOOR PLAN

VVDB 住宅，dmvA，比利时，梅赫伦，2009年

建筑师Van den Bergh要求dmvA事务所为其女儿翻新房屋。无需很多显著的改动，但需要一些微妙的修改，在地板上做些圆形的孔洞，设计一个新的中央螺旋楼梯并粉刷地板。一层设有入口以及主人的工作室。顶层有一个图书室，中间层设有起居室，厨房和卧室。所有的楼层都由红色的螺旋楼梯连接，显得更有活力和朝气。

House VVDB, dmvA, Mechelen, Belgium, 2009

Architect Van den Berghe requested dmvA to refurbish his own house for his daughter. No spectacular alterations, but subtle interventions, round perforations through the floors, a new central spiral stair, "whitening" of the floors. Located on the ground floor is the main entrance and the owner's studio. On the top floor it is the library. The levels in between consist of the living room, kitchen and bedrooms. All stories are connected by the red spiral stairs, adding more vigor to the interior.

材料：木材、铝件、水泥石

Materials: wood, aluminum joinery, cement stone

© Frederik Vercruysse

魅惑夜色
GLAMOROUS NIGHT

 小麦色 Wheat

 橙色 Orange

 暗金菊黄 Dark Goldenrod

 亮蓝灰 Light Slate Gray

 灰石色 Slate Gray

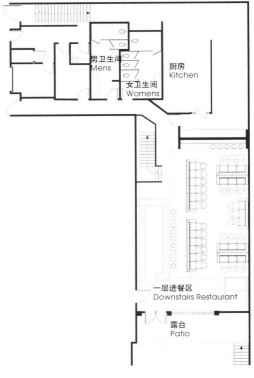

一层平面图 FIRST FLOOR PLAN

© Courtesy of Mister Important Design

© Courtesy of Mister Important Design

Motif 餐厅&夜总会，Mister Important Design，美国，加利福尼亚州，圣何塞市，2007年

　　Motif是一所独一无二的俱乐部，它将前卫的内饰和尖端的科技元素完美融合在一起。建筑的每一个角落都会带给游人不同的惊喜，如彩色灯光渲染出的单色基调、约12 m高的大幅花饰垂帘、由7 000片黑色玻璃构成的约30 m高的悬挂雕塑、光洁的白色乙烯基舞池地板，以及带有整体天窗的法式吊顶。

Motif Restaurant & Nightclub, Mister Important Design, San Jose, California, USA, 2007

Motif is a one of a kind venue that combines and avant garde interior with cutting edge technological features. There is a surprise for you around every corner with its' monochromatic palette lit with polychromatic lights, over-scaled patterned chain curtain installation, 40' suspended sculpture of 7,000 pieces of black glass, over 100 ft. of thick black rope curtaining, glossy white stretched vinyl dance floor and a French ceiling with integral blisters.

材料：花式垂帘、玻璃雕塑、乙烯基地板

Materials: fancy Chuilian, glass sculpture, vinyl flooring

© Courtesy of Mister Important Design

 亮粉红 Light Pink

 中紫罗兰红 Medium Violet Red

 蓝紫罗兰 Blue Violet

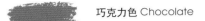 巧克力色 Chocolate

 纯红 Red

 猩红 Crimson

二层平面图 SECOND FLOOR PLAN

© Courtesy of Mister Important Design

© Courtesy of Mister Important Design

Motif 餐厅&夜总会，Mister Important Design，美国，加利福尼亚州，圣何塞市，2007年

俱乐部的内部空间显现出了现今最著名、最年轻有为的艺术家的风范。这三片区域，共两层的空间笼罩在LED艺术灯光以及各种传统和怀旧的元素之中。餐厅氤氲着惊艳的现代风格和静谧的浪漫气息，而楼上的酒吧则活力四射且舒适安逸。两个楼层均设有宽敞的天井卡座，Motif因而拥有了旧金山湾区最炫的室内和室外空间。

Motif Restaurant & Nightclub, Mister Important Design, San Jose, California, USA, 2007

Motif's interior represents some of today's most notable up and coming artists. The three-area, two-story space is lit with state of the art LED. lighting as well as traditional and retro lighting elements. The dining room is both startling modern and quietly romantic, whereas the upstairs lounge is both energetic and comfortable. With ample patio seating on both levels, Motif is truly one of the Bay Area's most striking indoor/outdoor spaces.

材料：花式垂帘、玻璃雕塑、乙烯基地板

Materials: fancy Chuilian, glass sculpture, vinyl flooring

© Courtesy of Mister Important Design

© Courtesy of Mister Important Design

迷幻之夜
PSYCHEDELIC NIGHT

 金色 Gold

 橙红色 Orange Red

 纯红 Red

 紫罗兰 Violet

© Courtesy of Mister Important Design

© Courtesy of Mister Important Design

210 North俱乐部，Mister Important Design，美国，内华达州，雷诺市，2009年

休息区是奢华的缩影，设有闪亮的铬质窗帘、天鹅绒的双人沙发和一个白色的大理石吧台。休息区最抢眼之处是由伦敦艺术家Eva Menz制作的吹制玻璃枝形玻璃吊灯。吊灯像一簇形态自由的3D立体波浪。它由5 000片玻璃构成，每一片玻璃都是由手工串起来的。吊灯反射着4个Elation品牌的Vision Color 250换色器射出的灯光，为休息区增添了光色变幻的体验。

210 North, Mister Important Design, Reno, Nevada, USA, 2009

The Divinity lounge is the epitome of luxury, with glittering chrome curtains, velvet loveseats, and a white marble bar. But its most magnificent feature is a blown-glass chandelier by London artist Eva Menz, shaped like a free-form 3D wave. The chandelier comprises 5,000 individual glass pieces, each of which had to be hand-strung. Four Elation Vision Color 250 color changers reflect light off of the chandelier, adding to the venue's color-morphing experience.

材料：铬质窗帘、天鹅绒沙发、大理石吧台、玻璃吊灯

Materials: chrome curtains, velvet sofas, marble bar, glass chandeliers

© Courtesy of Mister Important Design

	桃肉色 Peach Puff
	橙色 Orange
	橙红色 Orange Red
	蓝紫罗兰 Blue Violet
	深粉红 Deep Pink
	纯蓝 Blue
	纯红 Red

210 North俱乐部，Mister Important Design，美国，内华达州，雷诺市，2009年

俱乐部二层为人们与灯光的趣味互动提供了机会。这一空间原是一个废弃了的餐厅，在改造过程中也在技术和设计上向设计师提出了严峻的挑战。譬如，把曾经的扶梯改造成充满戏剧元素的好莱坞式入口再适合不过。设计者希望为LED灯装上复古的魅惑灯罩，使门廊和楼梯沐浴在逐渐变化的灯光之中。这一区域的吊顶极高，悬挂任何装置都是不可能的。

210 North, Mister Important Design, Reno, Nevada, USA, 2009

The second-floor club provided intriguing possibilities to do some really spectacular and different things with lighting. The space, a former restaurant that was completely gutted, also posed some serious technical and design challenges. For example, there was the escalator - ideal for staging a theatrical Hollywood-type entranceway - where the designers envisioned enchanting vintage-inspired lampshades wrapped around LEDs that would bathe the vestibule and stairwell in slowly changing colors. The massive height of the ceiling, however, made it virtually impossible to hang fixtures in the area.

材料：铬质窗帘、天鹅绒沙发、大理石吧台、玻璃吊灯

Materials: chrome curtains, velvet sofas, marble bar, glass chandeliers

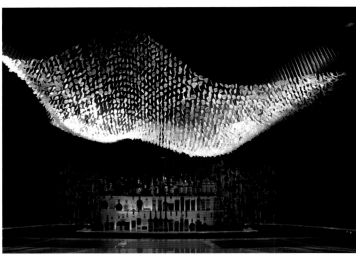

© Courtesy of Mister Important Design

© Courtesy of Mister Important Design

© Courtesy of Mister Important Design

© Courtesy of Mister Important Design

© Courtesy of Mister Important Design

© Courtesy of Mister Important Design

吉卜赛韵味
GYPSY FLAVOR

 金色 Gold

 巧克力色 Chocolate

 猩红 Crimson

 暗兰花紫 Dark Orchid

 中蓝色 Medium Blue

Gitane餐厅，Mister Important Design，美国，旧金山，2010年

"Gypsy"（吉卜赛）一词缘于希腊，西班牙语为"Gitano"，法语为"Gitan"。Gitane餐厅通过代表吉卜赛文化、音乐和美食的一系列丰富色彩、特色质地、各式物品将30年来的故事娓娓道来。餐厅的设计融入了吉卜赛传统，室内铺设了一系列花式繁复的织物、手工印花壁纸、复古吊灯及古董手工制品等。室内的装饰和质地为顾客创造了一个温暖、温馨的环境。酒吧的后墙挂着宽大的挂毯，各房间之间挂着大型帷帐，不仅营造出一种柔和的情调，同时在用餐人多时起到吸声作用。餐厅成功运用特色的质地、动感和纹饰，形成非凡的效果。另外，抬眼望去就看见餐桌上方那挂在金属板上的传统织物。用餐区以压低的顶部、柔软的织物和温暖的石墙形成一个温馨的用餐体验。

Gitane, Mister Important Design, San Francisco, USA, 2010

The word Gypsy derives from Egyptian, similarly to the Spanish Gitano or the French Gitan. Gitane surely delivers the story of three decades through an extensive palette of color, textures and objects that narrate the complexity of the Gypsy culture, music and gastronomic composition. Drawing from Gypsies historical accounts, the interior layers an array of intricate fabrics, hand printed wallpapers, vintage chandeliers and antique artifacts. The interior is filled with ornamentation and texture as a warm and inviting backdrop to patrons. Large tapestries are hung at the back wall of the bar; this, in addition to the large room-dividing draperies, softens the ambiance while simultaneously functioning to absorb the often loud dinner crowd. The restaurant's success of excessive texture, movement and ornamentation is extraordinary; soaring high above the dinner table, your eye follows the gathered fabric resting against the shiny planks of metal. The dining area is inviting as its lowered ceiling, soft fabrics and warm stone walls form an intimate dining experience.

材料：织物、壁纸、吊灯、挂饰、石墙

Materials: fabric, wallpaper, chandeliers, ornaments, stone walls

© Courtesy of Mister Important Design

© Courtesy of Mister Important Design

© Courtesy of Mister Important Design

热辣魅惑的空间
A HOT CHARM OF SPACE

 纯红 Red

 橙红色 Orange Red

 暗金菊黄 Dark Goldenrod

 黄土赭色 Sienna

 深粉红 Deep Pink

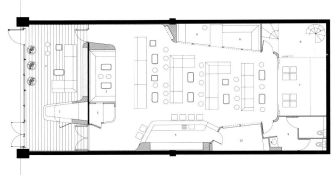

1 招待区	1 Reception
2 室内平台	2 Inside Terrace
3 贵宾室	3 VIP Room
4 酒吧	4 Bar
5 DJ间	5 DJ Booth
6 大腿舞房间	6 Lap Dance Room
7 舞台	7 Stage
8 后台	8 Backstage
9 卫生间	9 Toilets
10 储藏间	10 Storeroom

二层平面图 SECOND FLOOR PLAN

© Stéphane Chalmeau

© Stéphane Chalmeau

21号套房俱乐部，Trust in Design，法国，南特市，2007年

设计的灵感来自于数码贴画和其他平面设计作品的蕾丝图案，融合了怀旧乡村酒吧和现代流行艺术风格。红色亮漆镂空缎带遍布了整个俱乐部空间（酒吧、贵宾休息室、DJ间、大腿舞房间）。缎带勾勒出大型休闲吧和室内平台的轮廓。人们可以坐在各式各样舒适的沙发上观看表演或是聊天。缎带图案就像一个巨型的壁画，具有多层寓意，如蕾丝图案、煽动性小故事和隐晦的笑话。这是一个能让朋友们一起放松娱乐的热辣空间。

La suite 21 Club, Trust in Design, Nantes, France, 2007

The design's inspiration comes from the lace patterns for the computer-cut stickers and all the graphic design work. The old time provincial's cabaret mixed with contemporary pop culture. The red lacquered and perforated ribbon wraps up all the club's functions (bar, VIP lounge, DJ booth, lap dance rooms). This ribbon draws a large lounge and an indoor terrace where the furniture is set up. The sofas allow various and comfortable positions the see the show and chat with friends. The graphic ribbon is like a large fresco with different levels of meaning, from the lace pattern to small provocative stories and hidden jokes. It's a really good and hot space for people to relax and happy together with friends.

材料：红色亮漆

Material: red lacquer

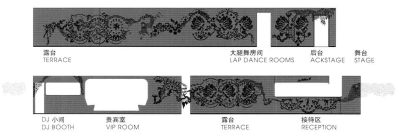

半透明贴纸 TRANSLUCENT STICKERS
穿孔图案 PERFORATIONS
黑色贴纸 BLACK STICKERS

立面图 ELEVATIONS

成长的力量
THE POWER OF GROWTH

 金色 Gold

 金菊黄 Goldenrod

 暗淡灰 Dim Gray

© Kim Myoung-sik

© Kim Myoung-sik

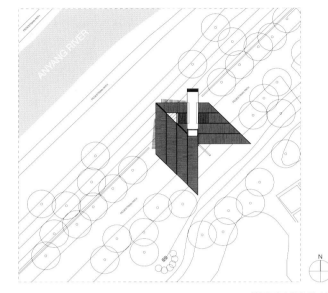
总平面图 SITE PLAN

立面图 ELEVATION

APAP开放学校，LOT-EK，韩国，安阳，2010年

8个船运集装箱沿45°角堆放并切割，形成了鱼骨状的结构，创造出箭头一样的体量，并且从地面抬升了3 m。建筑沿着Hakwoon公园的人行道布局，使该地成为聚集、休息及观景的中心地。建筑的颜色为明快的黄色与黑色，再加上印字和露天平台，十分突出，使这座建筑成为安阳市的地标和视觉焦点，车辆和行人远远地就可以看到。

APAP Open School, LOT-EK, Anyang, South Korea, 2010

Eight shipping containers are skewed to a 45 degree angle and combined in a fishbone pattern generating a large arrow-like volume that hovers three meters over the landscape. The structure is strategically placed over Hakwoon park pedestrian walkway at the city level right on the edge of the drop to the river bank, marking the territory as a focal place of gathering, resting and viewing. The strong graphic treatment of the new structure of the APAP Open School, with its bright yellow and black structure, lettering and deck, makes it visible for cars and passersby alike as a landmark within the urban fabric of Anyang.

材料：考顿钢

Material: Corten steel

© Kim Myoung-sik

奇妙之地
A PLACE OF WONDERS

 玫瑰棕色 Rosy Brown

 茶色 Tan

 印度红 Indian Red

卡瓦罗幼儿园，C+S Associati，意大利，特雷维索，2006年

  新建筑完美地融入了周围的景观。一面粗糙的混凝土墙涂以颜色，与周围的景致相匹配。墙上开有裂缝。因朝向不同，可随着光线变化而展现不同的效果。这栋建筑的核心即是它的结构：墙。墙，回缩或折叠，用颜色强调着它的通道、它的门廊。

  建筑的悬垂部分、稳固的砾石铺面以及光线都将门廊的视域拓宽，将室内的空间朝室外延伸，伴着声音和芬芳，将花园引入幼儿园来。这是对"门"的广义应用，将入口处转变为一个实际空间，将现实世界和理想世界联系起来，它位于室内和室外之间，特别且奇特，还伴随着犹豫、渴望、潜能与奇妙。

Nursery School Covolo, C+S Associati, Treviso, Italy, 2006

  The new building embraces the landscape. A rough concrete wall is colored to match the surrounding landscape, treated with split aggregate to reflect light in a variable manner depending on its orientation. The building is its structure: a wall. A wall that retracts and doubles, coloring itself to emphasize its passages, its thresholds.

  The overhang, the stabilized Sarone gravel paving and the lighting all expand the moment of the threshold, amplifying the classrooms spaces towards the exterior, or bringing the garden, with its sounds and scents, into the school. This operation of extension of the "door", this transformation of the threshold into an actual space become the imagination of a possible world, different and strange, suspended between inside and out. It represents hesitation, desire, potential, wonder.

材料：混凝土

Material: concrete

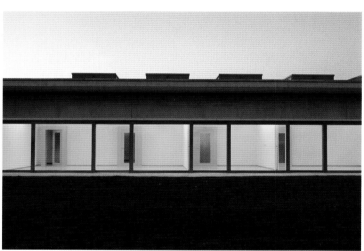
© Courtesy of C+S Associati

© Courtesy of C+S Associati

© Courtesy of C+S Associati

© Courtesy of C+S Associati

红与白融合激情与浪漫
RED AND WHITE BLEND PASSION AND ROMANCE

雪白色 Snow

猩红 Crimson

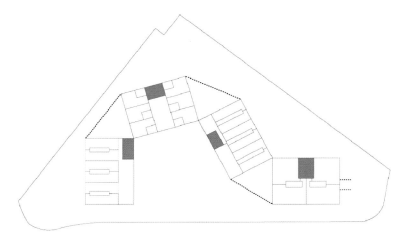

总平面图 SITE PLAN

© Adrià Goula

甘迪亚大学公寓，Guallart Architects，西班牙，瓦伦西亚，甘迪亚，2011年

  该项目旨在开发一座混合型建筑，使之符合相关的标准和特点，既可用做学生公寓，又能满足社会住房的需求。这一项目包括102间青年公寓、40间老年公寓和一个为镇议会而设的市政中心。该项目的另一重要特色是其绚丽的立面，红白相间，鲜艳夺目，象征着年轻学生居住者的激情与青春。

University Housing, Gandía, Guallart Architects, Gandía, Valencia, Spain, 2011

  This project aims to develop a hybrid project that would function essentially as a student residence while meeting the requirements of social housing, with the corresponding standards and characteristics. The program includes 102 apartments for young people, 40 apartments for senior citizens, and a civic and social center for the town council. Beyond the general layout, what's the most captivating about this project is the dramatic façade. The red and white cladding is quite overwhelming, representing the passion and youth of the young student residents in the building.

© Adrià Goula

材料：红、白覆面

Material: red and white cladding

秋天的童话
AUTUMN FAIRYTALE

 橙红色 Orange Red

 橙色 Orange

 茶色 Tan

 火砖色 Fire Brick

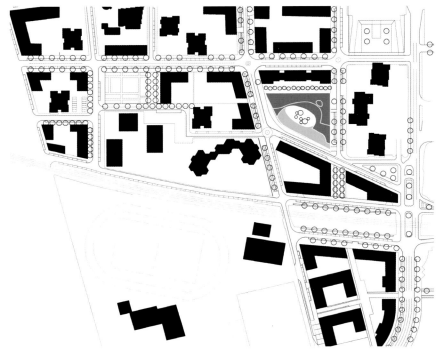

总平面图 SITE PLAN

© Florian Kleinefenn

博比尼学校，Mikou Design Studio，法国，博比尼，2012年

就像非洲，遍地都是阳光、野生动物以及宏伟的自然景色。那是一种神奇的力量，阳光的热度，大地的能量通通给设计师以灵感，带来一组洋溢着活力的橙红色系的拼合方案。

Complex School in Bobigny, Mikou Design Studio, Bobigny, France, 2012

It presents an image of Africa-sunshine, wild animal and grand natural sceneries everywhere. It's a magic power. Both the heat of the sun and energy of the earth can give designers the inspiration to work out a vigorous orange-red color scheme.

材料：天然木板、玻璃面板

Materials: natural timber panels, glass panels

橙红色 Orange Red

 橙色 Orange

 茶色 Tan

 火砖色 Fire Brick

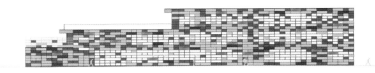

东立面图 EAST ELEVATION

南立面图 SOUTH ELEVATION

北立面图 NORTH ELEVATION

© Florian Kleinefenn

博比尼学校，Mikou Design Studio，法国，博比尼，2012年

秋天的一片落叶也许就是所有色彩的来源，记录花园里的黄色、棕色、红色及绿色如何相互融合，带来一组绝妙平衡的色彩。把这组颜色运用在相对低平、有着屋顶绿化的建筑中，再绝妙不过了。这样一来，建筑即是一个空中的花园，仿佛日日都享受着秋日的阳光。

Complex School in Bobigny, Mikou Design Studio, Bobigny, France, 2012

A falling autumn leaf can be the source of all the colors recording the combination of yellow, brown, red and green in the garden which is a group of perfectly balanced colors. It would be terrific if they were used for a relatively flat building with a green roof. The building is like a hanging garden which enjoys the autumn sunshine everyday.

材料：天然木板、玻璃面板

Materials: natural planks, glass panels

活力无限
ENDLESS FLOW OF ENERGY

 橙红色 Orange Red

 番茄红 Tomato

© Arne Quinze

© Arne Quinze

"重生"装置，Arne Quinze，法国，巴黎，2008年

"重生"是比利时设计师、艺术家Arne Quinze为巴黎著名的五星级酒店Le Royal Monceau的重新开张设计的木质雕塑装置，外形奇特，极具未来派特色。这个炫目的装置，只存在一个晚上，仿佛拥抱着整座酒店大楼，就像一条奔涌的溪流，狂野地穿破墙壁、走廊、楼梯、大厅和房间。长达15 km的荧光红木，辅以精妙的照明设施和带有40块LED屏的视频技术设施，这个经过精密测算的"重生"装置昭示了无穷无尽的活力，为人们带来了无比震撼的体验。"重生"装置将见证酒店的转型时刻，将揭示其动荡、历史和数不清的故事。它将酒店前几十年集聚的能量瞬时爆发了出来。

Installation Rebirth, Arne Quinze, Paris, France, 2008

"Rebirth" is a unique and futuristic wooden sculpture by Belgian designer and artist Arne Quinze for the well-known and distinguished five-star Paris hotel Le Royal Monceau in Paris. The dazzling sculptur-which lasted for only one exclusive night-embraced the entire hotel building as a conquering and boldness alien stream, breaking through walls, corridors, stairways, lobbies and rooms. With its 15 km of fluorescent red wood, refined illumination and 40 LCD-screens counting video installation, the carefully calculated endless flow of energy which "Rebirth" represents, offered the public a mind-blowing experience. "Rebirth" reflects the transition moment in which the building's DNA reveals its disorder, history and countless stories. It bursts out all its captured energy of the hotel's last decades.

材料：木材

Material: wood

发现之旅
DISCOVERY TRAVEL

 纯红 Red

 橙红色 Orange Red

 暗淡灰 Dim Gray

"旅行者"装置，Arne Quinze，德国，慕尼黑，2008年

"旅行者"，是Arne Quinze为庆祝路易威登在慕尼黑的新店开幕而设计的奇妙装置，是一个高20 m，宽12 m的木制建筑小品，于2008年11月在德国慕尼黑亮相。"旅行者"将带领游客开始一段奇妙的旅行，进入一个充满激情与灵感的世界。根据设计师的理念，旅行就意味着发现新的东西，目睹新的文化，并且经历新的生活体验。旅行即意味着灵感。

Installation The Traveller, Arne Quinze, Munich, Germany, 2008

On November 2008, Arne Quinze created a fantastic sculpture, a 20m high and 12m wide wooden architectural construction called "The Traveller" displayed at the occasion of the opening of the new Louis Vuitton store in Munich. Quinze invite us to a journey, an enigmatic travel through this masterpiece of which roads leads us to a world of emotion and inspiration. "Travelling means discovering new things, seeing new cultures, come across new aspects of life. I travel constantly and I consider it as enrichment for my evolution as a human being. Travelling equals to inspiration," declares Arne Quinze.

材料：木材

Material: wood

© Arne Quinze

© Arne Quinze

© Arne Quinze

架起沟通的桥梁
BRIDGING THE COMMUNICATION GAP BETWEEN PEOPLE

 橙红色 Orange Red

 硬木色 Burly Wood

"序列"装置，Arne Quinze，比利时，布鲁塞尔，2009年

"序列"装置是为2009年11月14至16日，在佛兰德议会大楼举办的政治节日而设计的，该装置将至少保留5年。装置为热烈的红色，由混凝土和木材建成。装置位于佛兰德议会与众议院之间，将二者连接起来，架起人们之间沟通的桥梁，并促进城市运动。这种物理上、象征意义上的连接体现了布鲁塞尔所有人都可能有着千丝万缕的联系。

The Sequence, Arne Quinze, Brussels, Belgium, 2009

This installation is built for the first Festival of Politics running on 14, 15 and 16 November in the Flemish Parliament, lasting for at least 5 years. It is all red, made of concrete and wood. It connects the Flemish Parliament with the House of Flemish Representatives, bridging the communication gap between people and generating movement in the city. The symbolic and physical connection between the neighbors, the Flemish Parliament and the House of Flemish Representatives, reflects a possible connection between all people in Brussels. Cross-culture connections, a connection with Europe, its diversity and entity.

材料：混凝土、木材

Materials: concrete, wood

© Arne Quinze

© Arne Quinze

© Arne Quinze

赤之热烈 | RED

木条搭建的奇幻世界
THE WOODEN STRUCTURES FANTASY WORLD

 橙红色 Orange Red

 珊瑚 Coral

雕塑作品，Arne Quinze，法国，巴黎，L'Eclaireur精品店，2009年

Sculptures, Arne Quinze, L'Eclaireur, Paris, France, 2009

材料：木材

Material: wood

© Arne Quinze

© Arne Quinze

© Arne Quinze

© Arne Quinze

索引
INDEX
412

住宅 RESIDENTIAL

光谱住宅
Spectrum Residential Ensemble —— 26

卡拉班切尔住宅楼
Carabanchel Housing —— 92

社会住房：63套公寓
Social Housing: 63 Apartments —— 102

2C住宅
2C Houses —— 120

Point Hyllie公寓楼
Point Hyllie —— 150

I' Park 城样板房
Model House for I' Park City —— 194

天井岛
Patio Island —— 214

NM别墅
Villa NM —— 218

克莱因瓶住宅
Klein Bottle House —— 222

织物立面公寓
Fabric Façade: Studio Apartment —— 252

西蒙娜公寓
The Simona —— 256

路易布兰科大街45号社会住宅
Social Housing in 45 Rue Louis Blanc —— 268

温伯格别墅
Haus am Weinberg —— 280

迪登家园
Didden Village —— 288

积分别墅
The Integral House —— 292

集装箱客房
Container Guest House —— 296

街角住宅
On the Corner —— 304

蓝色住宅
Blue Residential Tower —— 314

巴黎18区社会住宅
Social Housing in the 18th District of Paris —— 316

森拉住宅
Senra's House —— 336

东村公寓
The East Village —— 368

VVDB 住宅
House VVDB —— 380

教育 EDUCATION

MWD学院
Academie MWD —— 16

切萨皮克儿童发展中心
Chesapeake Child Development Center —— 28

可可幼儿园
Kindergarten Kekec —— 34

Los Mondragones日托中心和市政餐厅
Daycare Center and Municipal Dining Hall at Los Mondragones —— 38

马丁特小学
Martinet Primary School —— 40

蒙泰幼儿园
Monthey Kindergarten —— 46

维的亚兰卡尔理工学院学习中心X大厦 二期工程
'X' Block Vidyalankar Annex – Phase II —— 50

"青蛙王子"
Frog King —— 52

比尼萨莱姆学校综合建筑
Binissalem School Complex —— 56

贝里约扎尔幼儿园
Nursery School in Berriozar —— 76

南十字星圣母学校的图书馆与大厅
Our Lady of the Southern Cross College – Library & Hall —— 78

卡塔丽娜富兰克潘幼儿园
Kindergarten Katarina Frankopan —— 80

棕榈树丛中的幼儿园
Kindergarten Between Palms in Los Alcazares —— 84

法国皮卡尔幼儿园
Picard Nursey in Paris —— 106

荷兰代尔夫特理工大学学生公寓
Student Housing, the Technical University of Delft —— 108

中文	English	页码
研究实验室	Research laboratory	110
法国埃皮奈幼儿园	Epinay Nursery School	112
巴巴爸爸幼儿园	Kindergarten Barbapapà	116
幼儿中心	Infants' Center	122
康塞尔幼儿园	Consell Kindergarten	140
浩克兰德大学医院实验室	Laboratory Building at the Haukeland University Hospital	146
耶德鲁姆中学	Gjerdrum High School	152
波维尼尔小学扩建	Bovernier Primary School Extension	168
索夫盖德学校	Sølvgade School	200
MUMUTH音乐剧院	MUMUTH Music Theater	204
Braamcamp Freire中学	Braamcamp Freire Secondary School	232
贝纳通幼儿园	Day Care Center for Benetton	260
皮埃尔·布丁街托儿所	Crèche Rue Pierre Budin	264
平阳学校	Binh Duong School	274
雷蒙德幼儿园	The Leimond Nursery School	276
博索莱伊小学	Primary School in Beausoleil	282
圣伊莎贝尔幼儿园	Nursery School Santa Isabel	300
潘普洛纳幼儿园	Nursery School in Pamplona	330
塞露幼儿园	Kindergarten Selo	340
佛萨伦噶幼儿园	Nursery School in Fossalunga	344
沃戈斯克利中学扩建	Vogaskoli Secondary School Extension	346
奥胡斯工程学院	The New IHA Engineering College of Aarhus	350
诺凡西亚商学院	Novancia Business School	358
埃皮奈学生公寓	Student Housing in Epinay	360
意大利庞萨诺小学	Ponzano Primary School	362
APAP开放学校	APAP Open School	394
卡瓦罗幼儿园	Nursery School Covolo	396
甘迪亚大学公寓	University Housing, Gandía	398
博比尼学校	Complex School in Bobigny	400

办公 OFFICE

中文	English	页码
中川化工CS设计中心	Nakagawa Chemical CS Design Center	74
Infrax 办公楼	Offices Infrax West	126
Pita and Tecnova总部大楼	Pita and Tecnova Headquarters	130
IDOM总部大楼	IDOM Headquarters	134
巴乔马丁市政厅	Bajo Martín County Seat	248
法国电力集团总部的改造	Restructuration du siège EDF Ajaccio	254
SIC物流网络办公室	Logistics Complex of SIC Network	310
耶瑟奈斯行政中心	Administrative Center Jesenice	326
咨询楼	Advice House	354

商业综合体 COMMERCIAL COMPLEX

克林锦湖文化综合大厦
Kring Kumho Culture Complex — 182

格洛斯楚普购物中心
Glostrup Storcenter — 198

索里亚商场
Soria.com — 208

天安Galleria百货商场
Galleria Centercity — 302

升禧艺廊购物中心
Starhill Gallery — 318

酒店及餐饮 HOTEL & RESTAURANT

雅乐轩伦敦埃克塞尔酒店
Aloft London Excel — 20

阿布扎比洛克福特酒店
Rocco Forte Hotel Abu Dhabi — 144

盛开(餐厅室内设计)
Blossom (Interior Design of a Restaurant) — 246

波士顿大学学生公寓的"断层"咖啡厅
Fractal Café in Boston University Student Housing — 366

Gitane餐厅
Gitane — 390

休闲及娱乐 LEISURE & ENTERTAINMENT

EL "B" 礼堂
EL "B" Auditorium — 12

B.institut美容院
B.institut Beauty Parlor — 100

Masrah Al Qasba剧院
Masrah Al Qasba Theater — 216

流行音乐厅
The Sovereign Pop Venue — 226

Tschuggen Bergoase温泉中心
Tschuggen Bergoase Wellness Center — 306

Agora 剧院
Theater Agora — 372

Motif 餐厅&夜总会
Motif Restaurant & Nightclub — 382

210 North俱乐部
210 North — 386

21号套房俱乐部
La suite 21 Club — 392

博物馆及展馆 MUSEUM & PAVILION

布兰德霍斯特博物馆
Brandhorst Museum — 24

Flockr临时展馆
Flockr Pavilon — 158

宝格丽展亭
Bulgari Pavilion — 160

保罗和露露希利亚德大学艺术博物馆
Paul and Lulu Hilliard University Art Museum — 172

"瓣"亭
"Ban" Pavilion — 174

阿德勒天文馆内的克拉克家族欢迎画廊
Clark Family Welcome Gallery at the Adler Planetarium — 176

伯纳姆展馆
Burnham Pavilion — 180

新阿姆斯特丹广场与展亭
New Amsterdam Plein & Pavilion — 186

当代犹太博物馆
Contemporary Jewish Museum — 228

松恩&菲尤拉讷艺术博物馆
Sogn & Fjordane Art Museum — 236

卡尔·亨宁·彼得森 & 埃尔斯·阿尔菲尔特博物馆
Carl-Henning Pedersen & Else Alfelt Museum — 312

贝诺佐·戈佐利博物馆
Benozzo Gozzoli Museum — 334

咖啡机博物馆
Museum MUMAC — 370

公共服务设施 PUBLIC SERVICE FACILITY

巢鸭信用银行江古田支行
Sugamo Shinkin Bank / Ekoda Branch —————— 58

巢鸭信用银行新座支行
Sugamo Shinkin Bank / Niiza Branch —————— 62

巢鸭信用银行志村支行
Sugamo Shinkin Bank / Shimura Branch —————— 66

巢鸭信用银行常盘台支行
Sugamo Shinkin Bank / Tokiwadai Branch —————— 70

Sa Indiotería 体育中心
Sports Center Sa Indiotería —————— 88

动物庇护所
Animal Refuge Center —————— 104

阿尔茨海默氏症患者的新住院设施
New Accommodation Facility for Alzheimer´s —————— 136

加瓦消防站
Fire Station in Gavà —————— 154

1号停车场
Car Park One —————— 190

2号停车场
Car Park Two —————— 192

流浪者之家
Shelter Home for the Homeless —————— 210

索莫茹社区中心
Sōmeru Community Center —————— 322

梅里达工厂青年活动中心
Merida Factory Youth Movement —————— 378

展览及装置 Exhibition & INSTALLATION

草间弥生作品
Yayoi Kusama´s Work —————— 94

中川化工CS设计中心（"万花筒"展览）
The Nakagawa Chemical CS Design Center, "Kaleidoscope" Exhibition —————— 164

更衣室
The Changing Room —————— 188

"空间演化"展览
Exhibition Evolution of Space —————— 212

亨里克·奥利维拉作品
Henrique Oliveira´s Work —————— 240

陶瓷云
CCCloud —————— 270

阿纳·奎兹作品
Arne Quinze´s Work —————— 404